Human Body, the Coloring Book

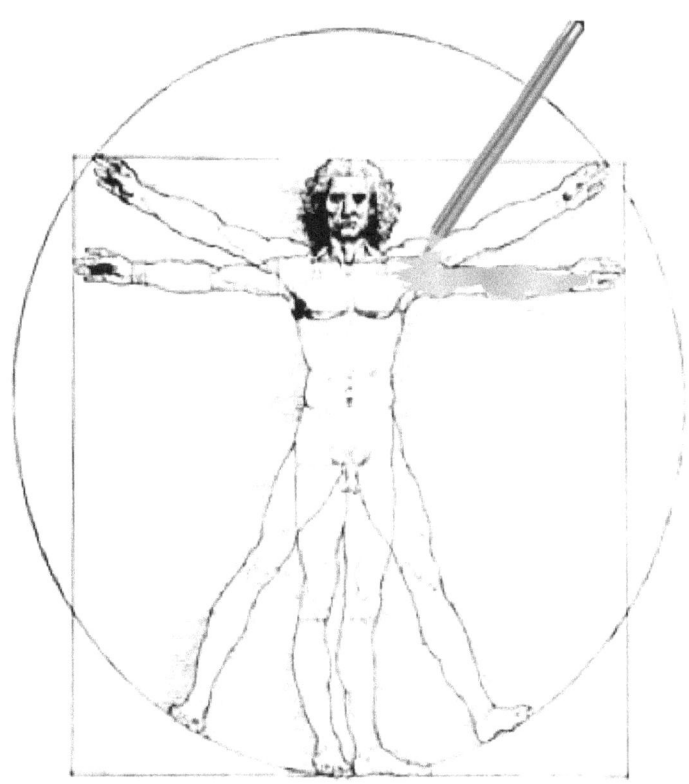

By James Clark

Human Body, The Coloring Book
© J.E. Clark, 2024. All rights reserved.

Published by James E Clark.
All rights reserved. No part of this publication may be reproduced or transmitted in any form or by any means, electronic or mechanical, including photocopy, recording, or any information storage and retrieval system, without permission.

ISBN: 9798324922047

PREFACE:
Human Body, the Coloring Book provides a comprehensive overview of human anatomy for those visually oriented and kinesthetic learners. This atlas is not a traditional textbook. It requires active input from the students ensuring they will learn and remember concepts much more effectively than with traditional textbooks alone. Completing coloring exercises can also serve as tools to review and prepare for examinations.

This book is suited for students beginning their study of the human body through those taking their first college course in human anatomy and physiology. Human Body, the coloring book is a valuable supplement to any anatomy and physiology text. With exercises that can also serve as stand along educational resources for the students, either within the classroom or as assigned homework.

Why Color?
Coloring is an excellent way for students to learn about the structure, anatomy, of their body. It will assist with the memorization that is necessary to learn anatomy. As the process of coloring helps students remember details and factual information, as they must pay attention to details, visualize structures, and may physically feel the relationship between different structures of the body as they color. Students using the Human Body, the Coloring Book will be able to deepen their understanding of human anatomy as they can visualize the participation of structures necessary for the systems, tissues and organs of their body to function. Best of all, coloring is fun for students—a welcome distraction from more static studying activities such as reading and memorizing!

Organization
Human Body, the Coloring Book follows the systems approach favored by traditional anatomy and physiology textbooks, so it can be used with any such book. Beginning with the fundamental concepts in anatomy, cell biology, and histology. Subsequent images deal with the anatomy of different body systems, and need not be completed in order.
The intent of the organization and coloring the structure name and the structure will help students remember the spelling and location of the structure. In addition, the completed diagram will serve as a useful reference and review tool, since it will be easy to match different structures to the different terms.

Table of Contents:

Image	Page #
Anatomical Terminology: How to speak about Planes, Places, Locations like a scientist	
Overview	1
Definition of Terms	2
Anatomy Identification	
1:Planes	3
2: Relationships	4
3: Locations	5
Anatomical Landmarks	6
1: Anterior Landmarks	7
2: Posterior Landmarks	8
Cells, Tissues, Organs: Levels of Organization the Human Body	
Overview	9
Levels of Organization	10
Cells and Organelles	11
Cell Membrane	
1: Structure	12
2: Transportation	13
Cells and Tissues	14
Systems of the Body	15
Integument System: Skin and structures of the skin	
Overview and Functions	16
Integument Anatomy	
Structures	17
Epidermis	18
Sensory Structures	19
Hair and Nail	20
Glands of the Body	21
Musculoskeletal System: Bones, Skeletal Muscles and Joints of the Body	
Skeletal Overview	22
Bone Structure	
1: Landmarks and Anatomy of the bone	23
2: Tissues	24
Articulations	
1: Joint Types	25
2: Synovial Joints	26
Boney Landmarks	27
Skeleton	
1: Overview	28
2: Cranial and Facial Bones	29
3: Vertebrae and Curvatures	30
4: Vertebrae Identification and Landmarks	31
5: Thoracic Cage and Costal Bone Identification and Landmarks	32
6: Upper Extremity Bones	33
7: Scapular and Clavicle	34
8: Upper Extremity Bones Identification and Landmarks 1 (Humerus)	35
8: Upper Extremity Bones Identification and	36

	Landmarks 2 (Ulna and Radius)	
	9: Hand and Wrist (Carpals and Metacarpals) Identification	37
	10: Lower Extremity Bones	38
	11: Pelvic Bones Identification and Landmarks	39
	12: Lower Extremity Bones Identification and Landmarks 1 (Femur)	40
	12: Lower Extremity Bones Identification and Landmarks 2 (Tibia and Fibula)	41
	13: Foot (Tarsals and Metatarsals) Identification	42
Skeletal Muscle Overview Structure		43
	1: Muscle Structure	44
	2: Tissues	45
	3: Neuromuscular Junction	46
Muscle		
	1: Overview (Anterior)	47
	2: Overview (Posterior)	48
	3: Head and Neck	49
	4: Upper Extremity	50
	5: Abdominal/ Trunk Muscle	51
	6: Lower Extremity (Anterior)	52
	7: Lower Extremity (Posterior)	53
Nervous System: Neurons, Nerves and Structures of the Cerebral Cortex		
	Overview	54
Tissues		
	1: Cells	55
	2: Synapse and Action Potentials	56
	3: Action Potential	57
Cerebral Cortex		
	1: Medial and Lateral View	58
	2: Coronal and Inferior View	59
	3: Ventricles and Cerebrospinal Fluid	60
Spinal Cord		
	1: Structures and Spinal Nerve	61
	2: Dermatome and Myotomes	62
Peripheral Nerves		
	1: Brachial Plexus	63
	2: Lumbar/Sacral Plexus	64
Special Senses		
	1: Olfaction and Gustation	65
	2: Auditory	66
	3: Vestibular	67
	4: Vision 1 (Structures)	68
	4: Vision 2	69
Sensation		
	1: Peripheral Sensory Receptors	70
	2: Somatosensory Cortex Topographic Organization	71
Neuroendocrine System: Hormones, Glands and Tissues involved with homeostasis		
	Overview	72
Neuroendocrine Tissues and Glands		

1: Tissues Overview	73
2: Pituitary	74
3: Thyroid and Parathyroid	75
4: Adrenal	76
5: Pancreas	77
6: Gonads	78
Cardiorespiratory System: Heart, Blood, Blood Vessels, Lungs and Gas Exchanges	
Cardiovascular Overview	79
Key Terms for the System	80
Cardiorespiratory Overview	81
Cardiovascular	
1: Vessel Pathway	82
2: Blood	83
3: Heart 1- External Anatomy	84
Heart Properties	85
3: Heart 2- Internal Anatomy	86
4: Heart Cycle	87
5: Circulation through Upper Extremities of the Body	88-89
6: Circulation through Lower Extremities of the Body	90-91
7: Circulation through the Thorax and Abdomen: Branches of Aorta and Vena Cava	92-93
8: Circulation through Cranium and Cerebral Cortex	94-95
Respiratory Overview	96
Respiratory Terms and Gas Laws	97
Respiratory	
1: Overview	98
2: Structures of the System	99-101
3: Alveoli	102
4: Ventilation Cycle	103-104
Gas Exchange	105
Immune System: Barriers, Immune Cells, Vaccination, the protection against infection and response to injury	
Overview	106-107
Lymph Nodes and Vessels	108-109
Cell Response	
T- Lymphocyte (Cell) Response	110
B-Lymphocyte (Cell) Response	111
Antibody (Immunoglobulin)	112
Allergen Response	113
Vaccination	
Overview	114
Herd Immunity	115
Gastrointestinal System	
Overview	116-117
System Organs	
1: Overview	118
2a: Oral Opening Structures	119
2b: Oral Opening Teeth	120
3: Alimentary Canal	121
4: Stomach	122
5: Small Intestine	123
6: Large Intestine	124
7: Liver	125

	8: Pancreas	126
Excretory System: Kidney & Renal Functions and Water Balance		
	Overview	127-128
	Renal System	
	1: Overview	129
	2: Renal Anatomy	130
	3: Nephron Anatomy	131
	Hydration and Water Balance	132
Reproductive System & Development		
	Overview	133-134
	Male System	135-137
	Female System	138-141
	Meiosis and Gametogenesis	142-143
	Fertilization and Implantation	144
	Embryogenesis	145
	Fetal Circulation	146
	Summary of Development	147

Anatomical Terminology

How to speak about
Planes, Places,
Locations like a scientist

Overview

Being a scientific discipline, the study of human anatomy and physiology is filled with terminology that has at its roots classical words, word fragments and meanings. These terms extend into the closely related disciplines within the medical fields. As such, it is expected that students of anatomy and physiology will be excessively exposed to these words and word fragments in their classical language forms (Greek and Latin). In particular, as they are extensively used to name structures and concepts so as to ensure these structures and concepts can be internationally understood during studying anatomy. Therefore, we need to take time to review the meanings behind the technical terms used in the study of anatomy and used by professionals across medical, science and health professions.

When discussing anatomical terms, the references that we use will depend on where you are attempting to reference. the development of three reference planes that generally used to divide the body, or regions of the body, about the imaginary geometric axis. These planes are the CORONAL (FRONTAL), that divides along the X-axis of the body, the TRANSVERSE (CROSS-SECTIONAL, TRANSAXIAL)t that divides along the Z-axis of the body, and the SAGITTAL, that divides along the Y-axis. There is a special case within the sagittal plane (MID-SAGITTAL) that divides evenly into two halves. There are also references as to the position of a structure relative to other parts of the body, or body segments. This type of reference is dependent upon the point of reference that is being taken in explaining the location of the structure of interest. With that in mind, we must discuss the anatomical terminology based on reference to the body as a whole, or based on a regional area of the body or body segment,

Definition of Terms

SAGITTAL: Plane that divides body, or region, into "RIGHT" and "LEFT"

CORONAL: Plane that divides body, or region, into "FRONT" and "BACK" sometimes referred to as **FRONTAL**

TRANSVERSE: Plane that divides body, or region, into "ABOVE" and "BELOW"

VENTRAL: Toward the Stomach side of the body (front of the body)

DORSAL: Toward the Vertebral side of the body (back of the body)

CRANIAL: Toward the head

CAUDAL: Toward the tail

IPSILATERAL: On the same side of the body

CONTRALATERAL: On the opposite side of the body

ANTERIOR: Toward the front side of the region, organ, or limb

POSTERIOR: Toward the back side of the region, organ, or limb

PROXIMAL: Closer to the torso

DISTAL: Further from the torso

LATERAL: Away from the midline of the long axis of the body or region of the body

MEDIAL: Closer to the midline of the long axis of the body or region of the body

SUPERIOR: Region is above along the long axis of the body or region of the body

INFERIOR: Region is below along the long axis of the body or region of the body

SUPERFICIAL: Closest to the outer surface

DEEP: Furthest from the outer surface

Anatomy Identification 1: Planes

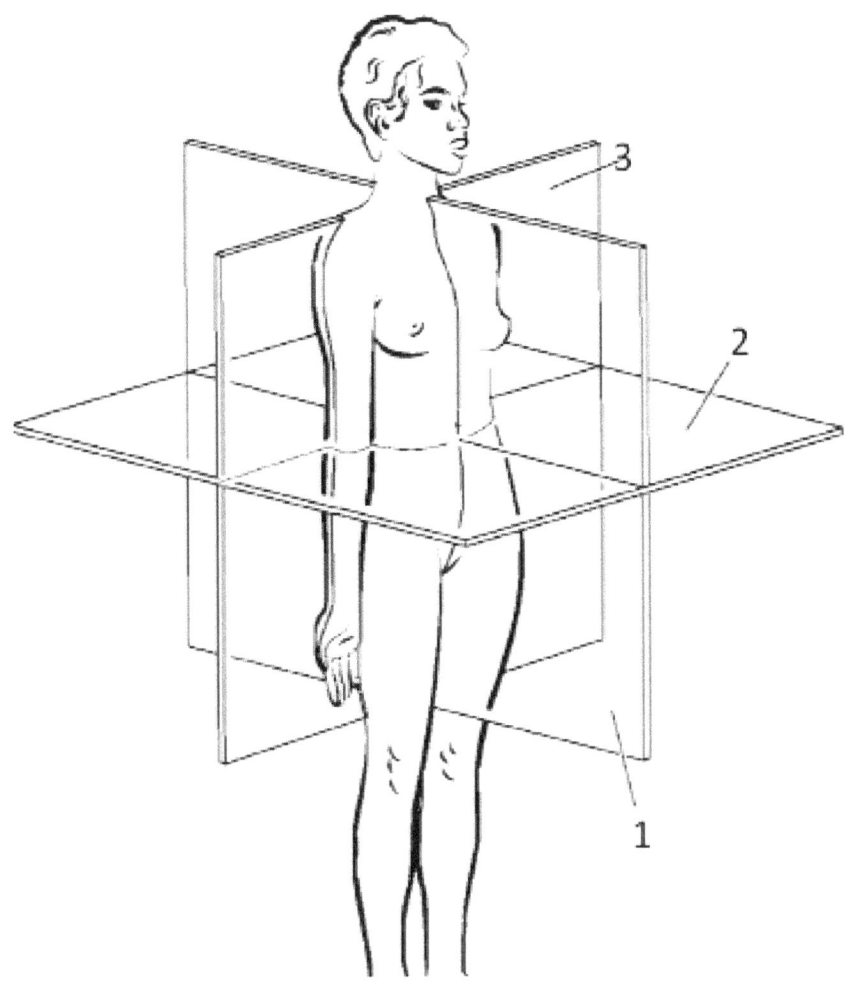

1 Sagittal
2 Transverse
3 Coronal

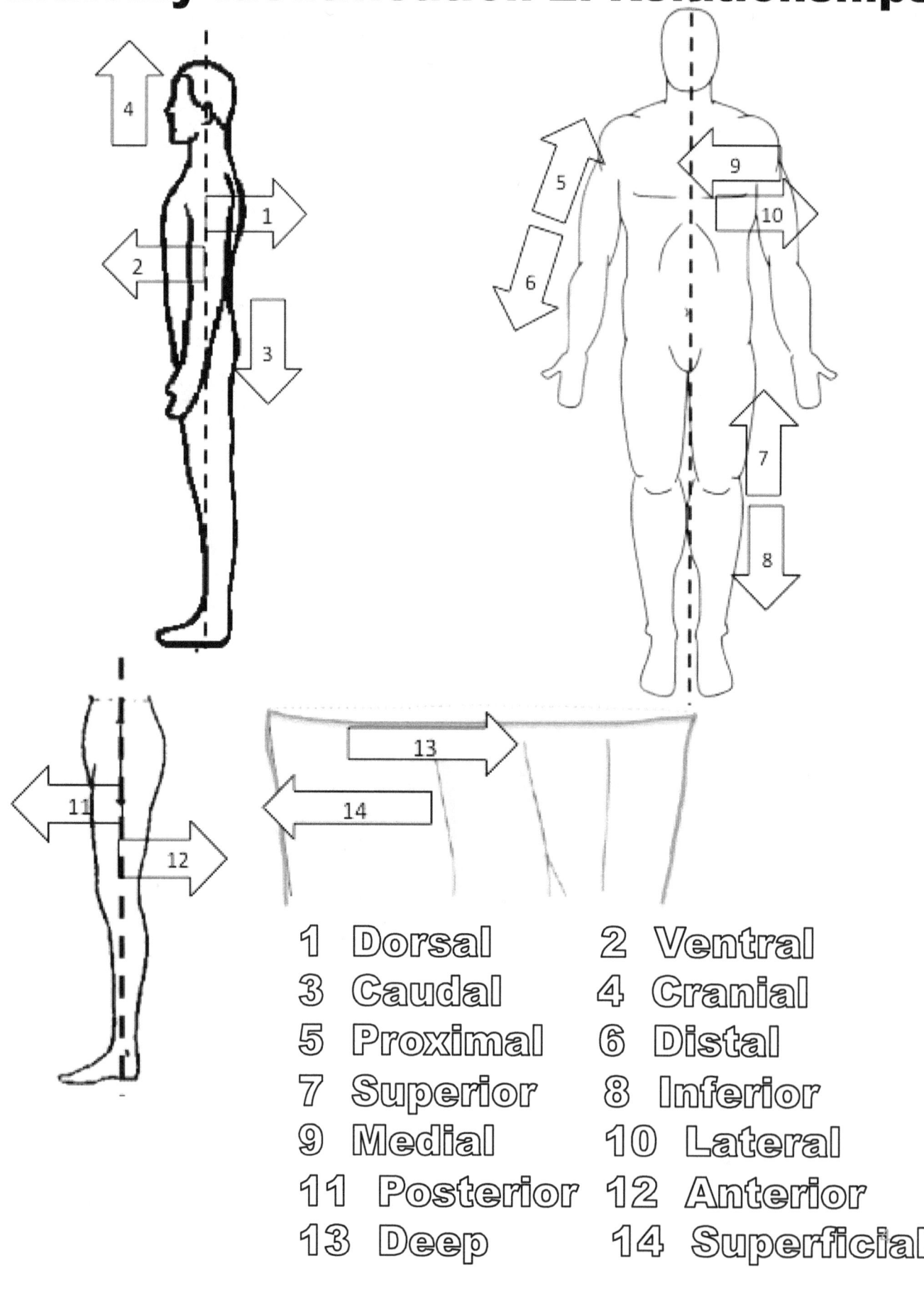

Anatomy Identification 3: Location

1=Ventral Cavities
2=Dorsal Cavities
3=Abdominopelvic Cavities
4=Cranial Cavity
5=Vertebral Cavity
6=Diaphragm
7=Thoracic Cavities
8=Abdominal Cavity
9=Pelvic Cavity
10=Right Hypochondriac
11=Epigastric
12=Left Hypochondriac
13=Right Lumbar
14=Umbilical
15=Left Lumbar
16=Right Iliac
17=Hypogastric
18=Left Iliac

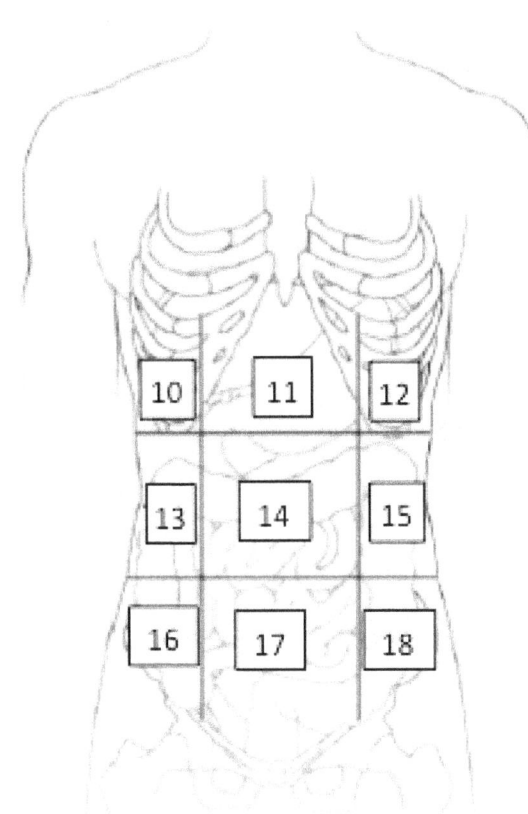

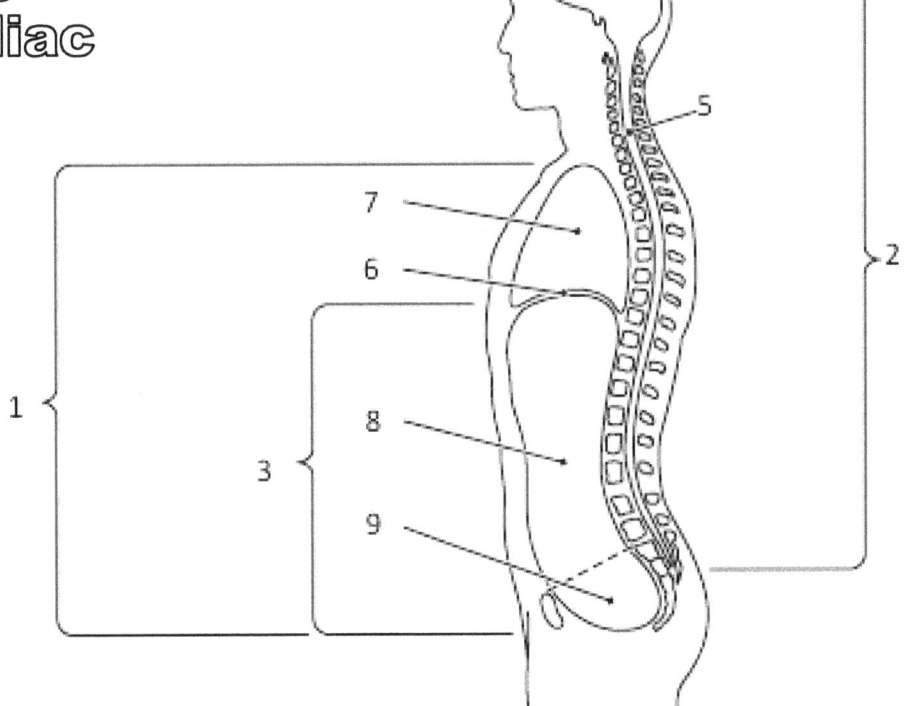

Anatomical Landmarks

Nose	Nasal	Chest	Pectoral
Mouth	Oral	Spinal Column	Vertebrae
Neck	Cervical	Low Back	Lumbar
Head	Cephalic	Tail Bone	Coccygeal
Head (Brain)	Cranial (Cranium)	Belly Button (Navel)	Umbilicus
Forehead	Frontal		
Eye	Orbital (Ocular)	Abdomen ("Stomach")	Abdominal
Cheek	Buccal		
Mental	Chin	Abdomen ("Lower Abdomen")	Pelvic
Ear	Aurical (Otic)		
Side of Head	Temporal		
Back of Head	Occipital	Gentiles	Pubic
Rib	Costal	Outside of Shin	Fibular
Sternum	Sternal	Area around Anus	Perineal
Breast	Mammary	Crotch	Inguinal

Shoulder	Deltoid/Acromial		
Armpit	Axillary	Kneecap	Patellar
Arm	Brachial	Back of Knee	Popliteal
Forearm	Antebrachial	Calf	Sural
Front of Elbow	Antecubital	Leg (shin)	Crural
Back of Elbow	Olecranon	Ankle	Tarsal
		Heel	Calcaneal
Wrist	Carpal	Foot	Pes
Hand	Manus	Top of foot	Dorsum
Palm of Hand	Palmar	Sole of foot	Plantar
Back of Hand	Dorsum	Big Toe	Hallux
Fingers/Toes	Digits	Thumb	Pollex (Polis)
Hip	Coxal	Collar Bone	Clavicle (Clavicular)
Buttock	Gluteal		
Thigh	Femoral		

Anatomical Landmarks 1: Anterior Landmarks & Terminology

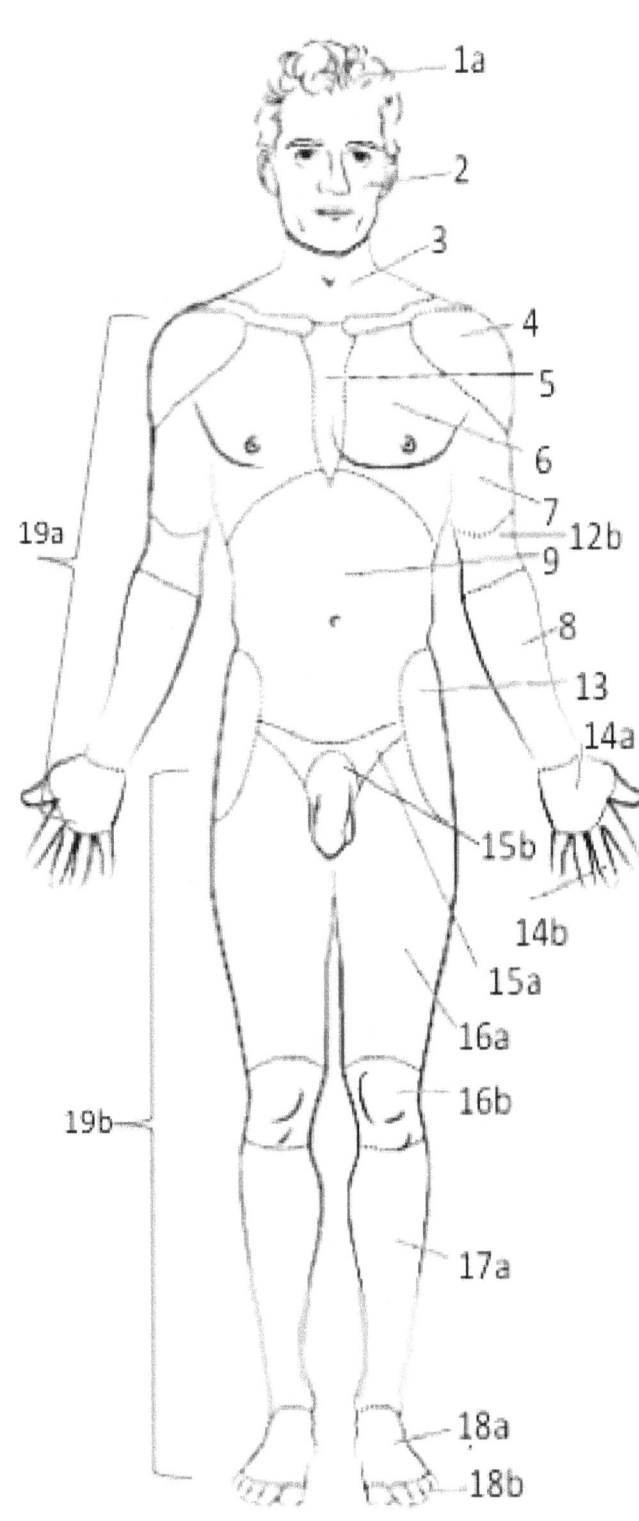

1a Cranial
2   Facial
3   Cervical
4   Deltoid
5   Sternal
6   Pectoral
7   Brachium
8   Antebrachium
9   Abdominal
12a Olecranon
12b Antecubital
13 Cox
14a Palmar of Hand
14b Digits of Hand
15a Inguinal
15b Pubic
16a Thigh
16b Patellar
17a Cural
18a Dorsum of Foot
18b Digits of Foot
19a Upper Extremity
19b Lower Extremity

Anatomical Landmarks 2: Posterior Landmarks & Terminology

1b Occipital
3   Cervical
4   Deltoid
7   Brachium
8   Antebrachium
10a Scapular
10b Vertebral
10c Lumbar
11 Gluteal
12a Olecranon
13 Cox
14b Digits of Hand
14c Dorsum of Hand
16a Thigh
16c Popliteal
17b Sural
18c Calcaneal
19a Upper Extremity
19b Lower Extremity

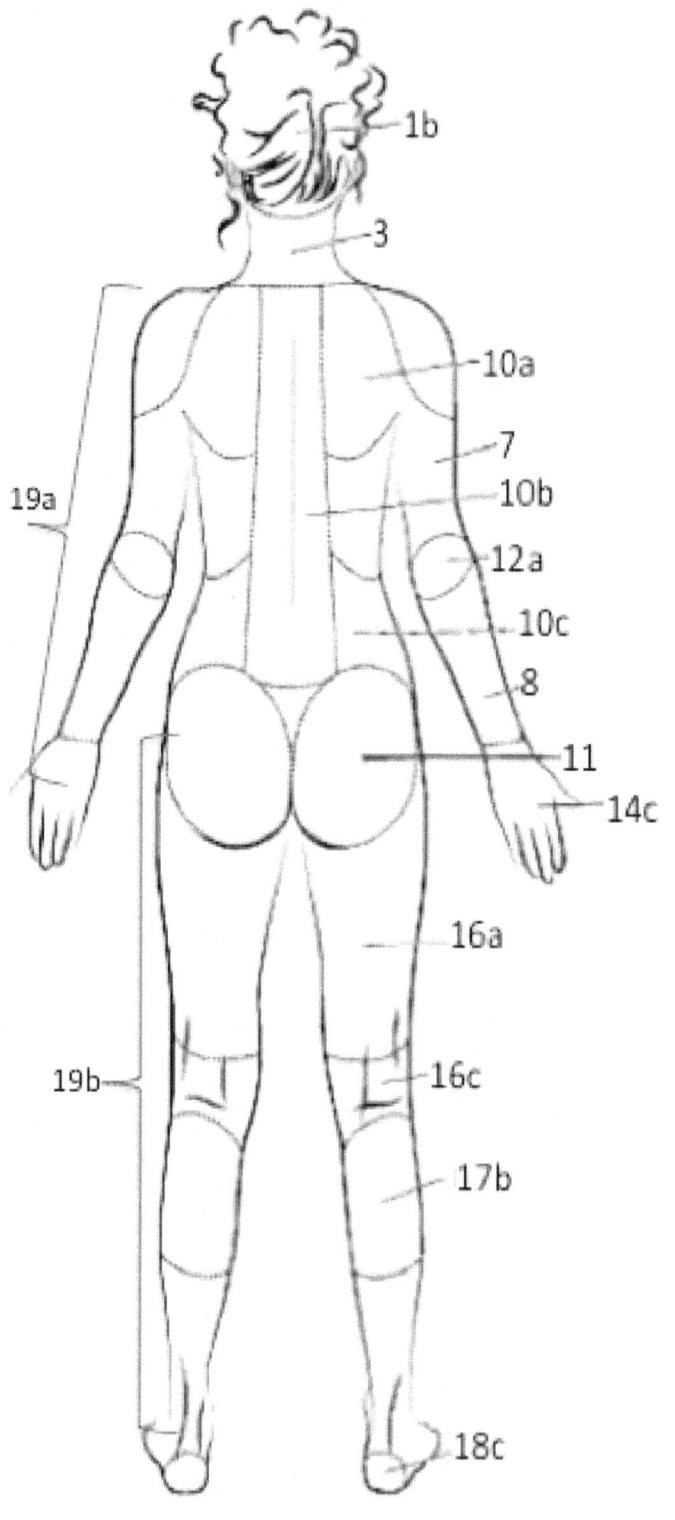

Cells, Tissues, Organs

Levels of Organization of the human body

Overview

The body is a complex network of cells that group together into distinct structures that allows for the body to function. In order to begin to understand this we must first look at how the body is built from the various building blocks eventually forming the body. In this, we must think of the body as being built in layers of ascending complexity beginning with the ATOM and MOLECULE and ending with entire HUMAN BODY. Each level of complexity is developed through an increase in the various components that are interacting within that level. Which begins with the atom and the subatomic components (electrons, neutrons, protons) followed by the interaction of atoms with other atoms (or molecules) and then single molecules with other molecules forming the macromolecules that we tend to think about in biology (e.g., carbohydrates, lipids, proteins). From these MACROMOLECULES we have interactions that eventually from the ORGANELLES and CELLS that will interact with each other leading to the formation of the tissues. TISSUES are conglomerations of cells that share a similar function for the body that will work and interact with each other. Developing into regulated (or control) ORGANS, a conglomeration of tissues with a shared function within the homeostasis of the body. These organs eventually coordinate their independent functions into the ORGAN SYSTEMS that comprise the body that we typically think about when discussing human anatomy and physiology.

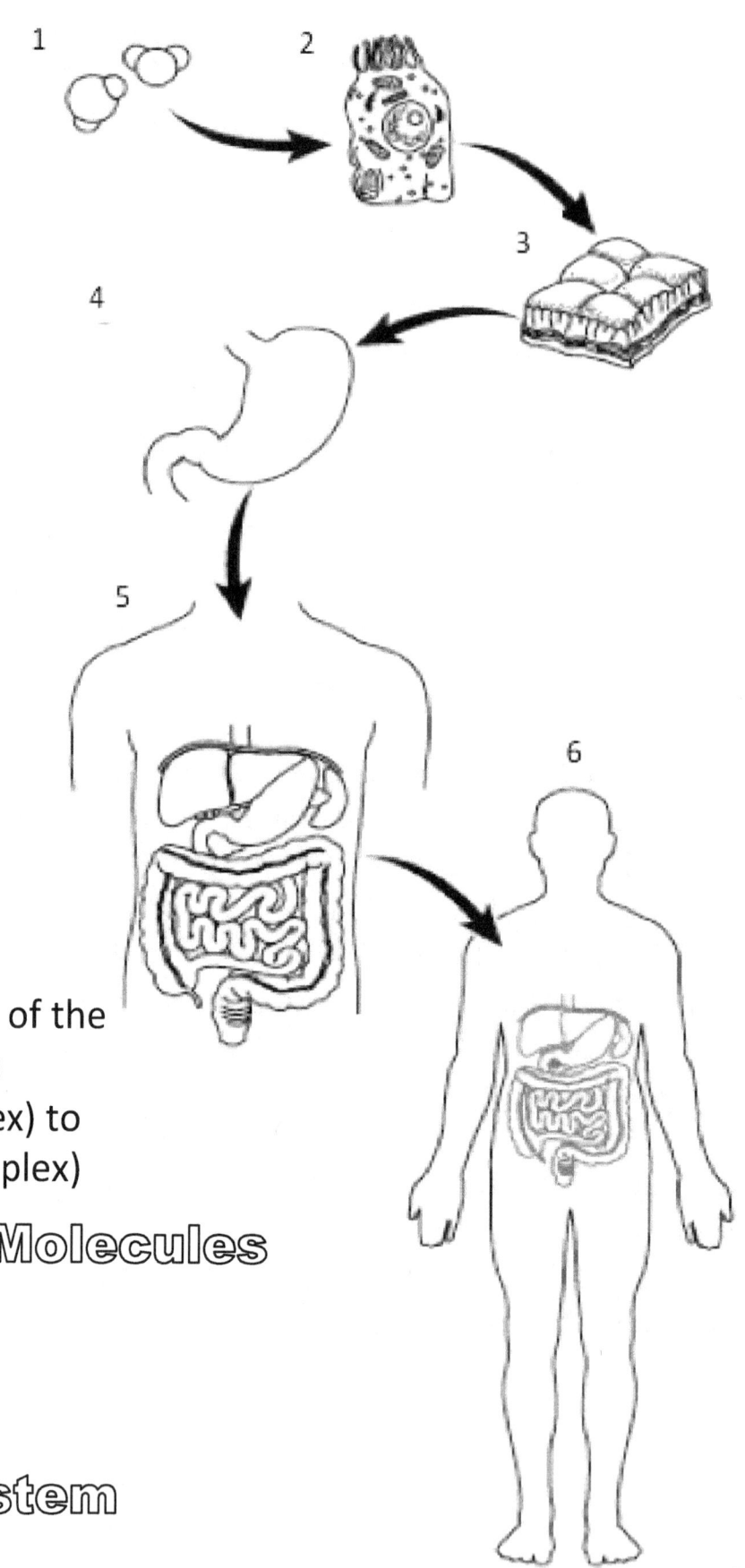

Levels of organization of the human body from the smallest (least complex) to the largest (most complex)

1. Atoms & Molecules
2. Cell
3. Tissues
4. Organs
5. Organ System
6. Organism

Cell and Organelles

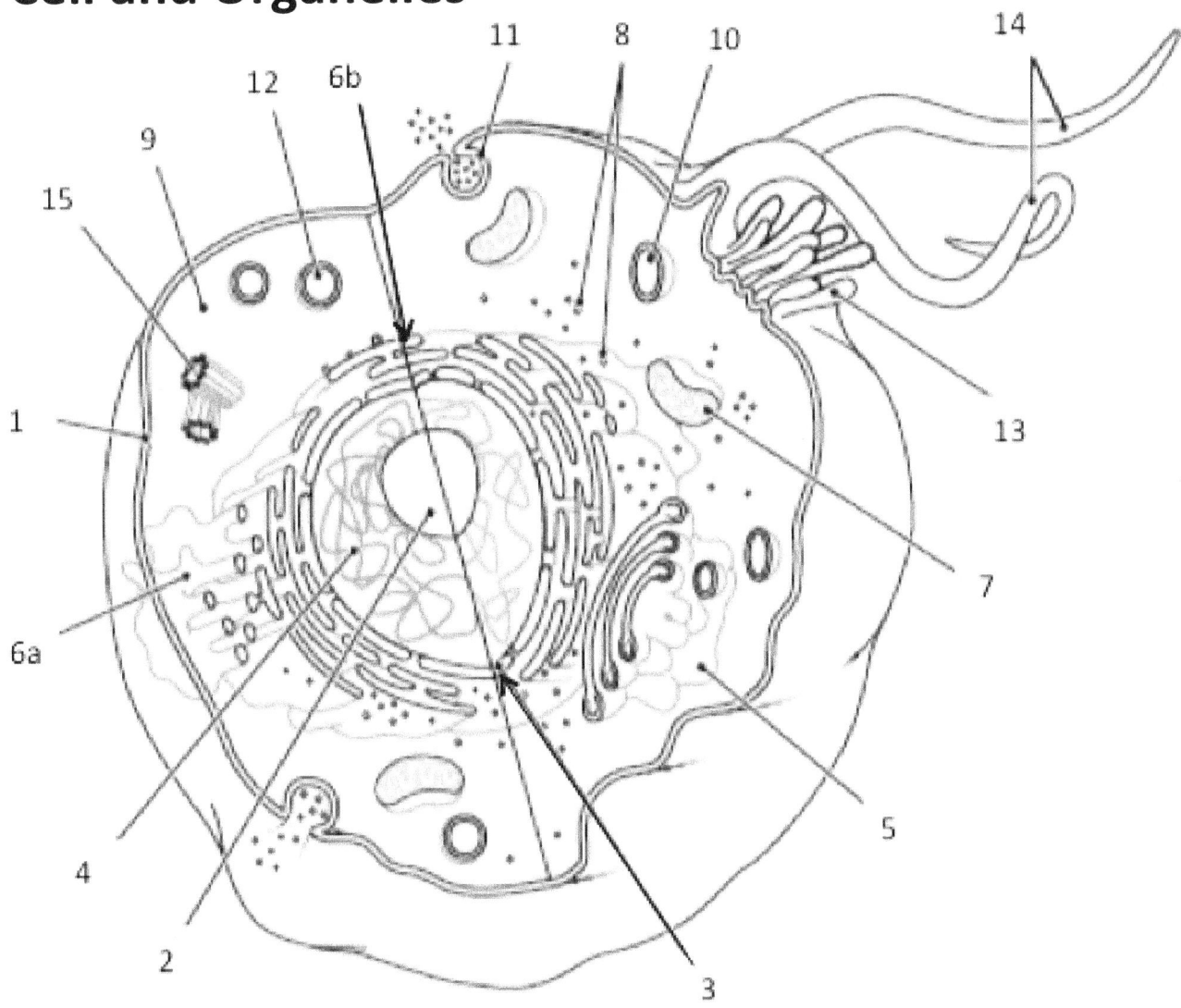

1. Cell Membrane
2. Nucleolus
3. Nuclear Envelope
4. Nucleus
5. Golgi Body/ Apparatus
6a. Smooth Endoplasmic Reticulum
6b. Rough Endoplasmic Reticulum
7. Mitochondria
8. Ribosomes
9. Cytoplasm
10. Lysosome
11. Exocytosis of Vacuole
12. Peroxisome
13. Cilia
14. Flagella
15. Centriole

Cell Membrane 1: Structure

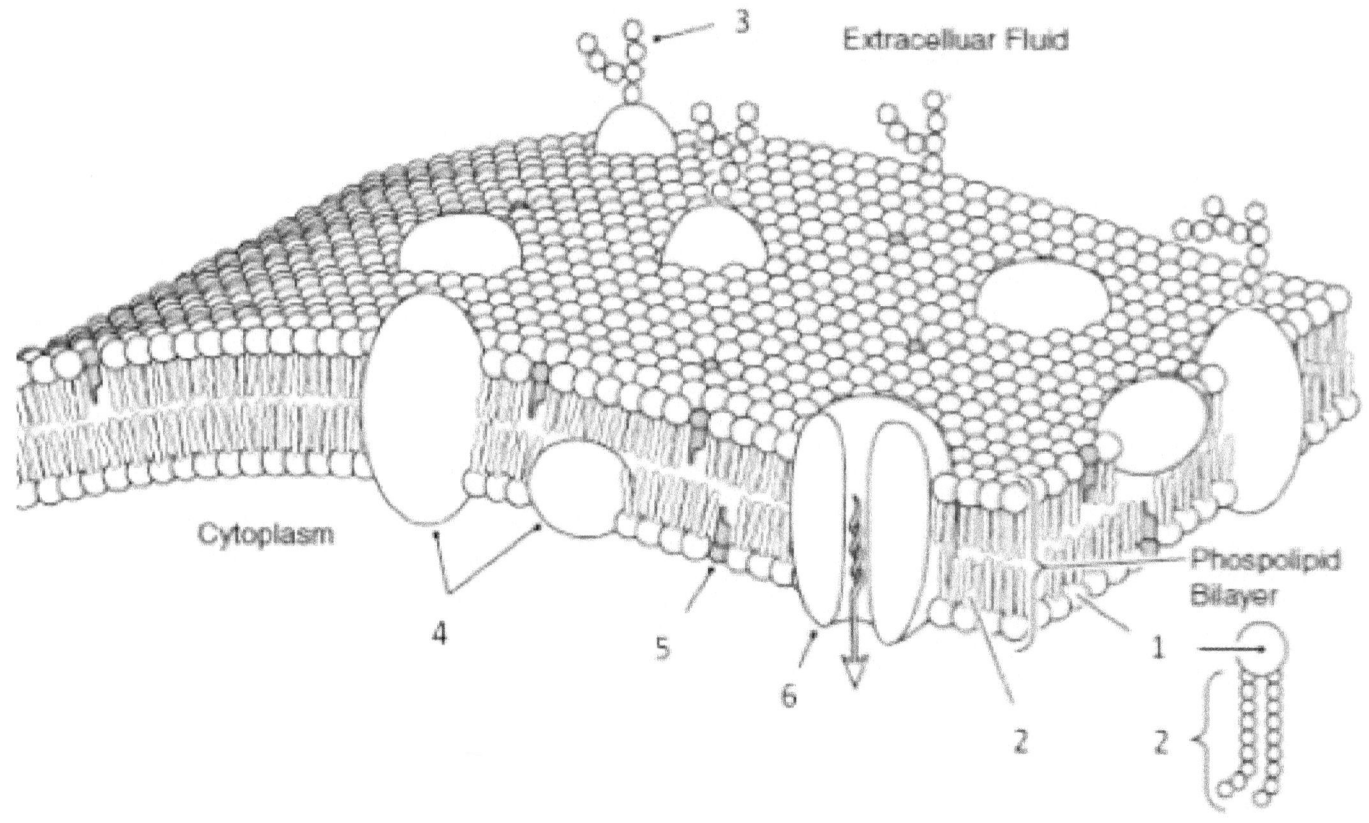

1. Hydrophilic Head of Phospholipid
2. Hydrophobic Tail of Phospholipid
3. Antigen Maker (Glycoprotein)
4. Membrane Protein
5. Cholesterol Molecule
6. Transmembrane Protein Channel

Cell Membrane 2: Transportation

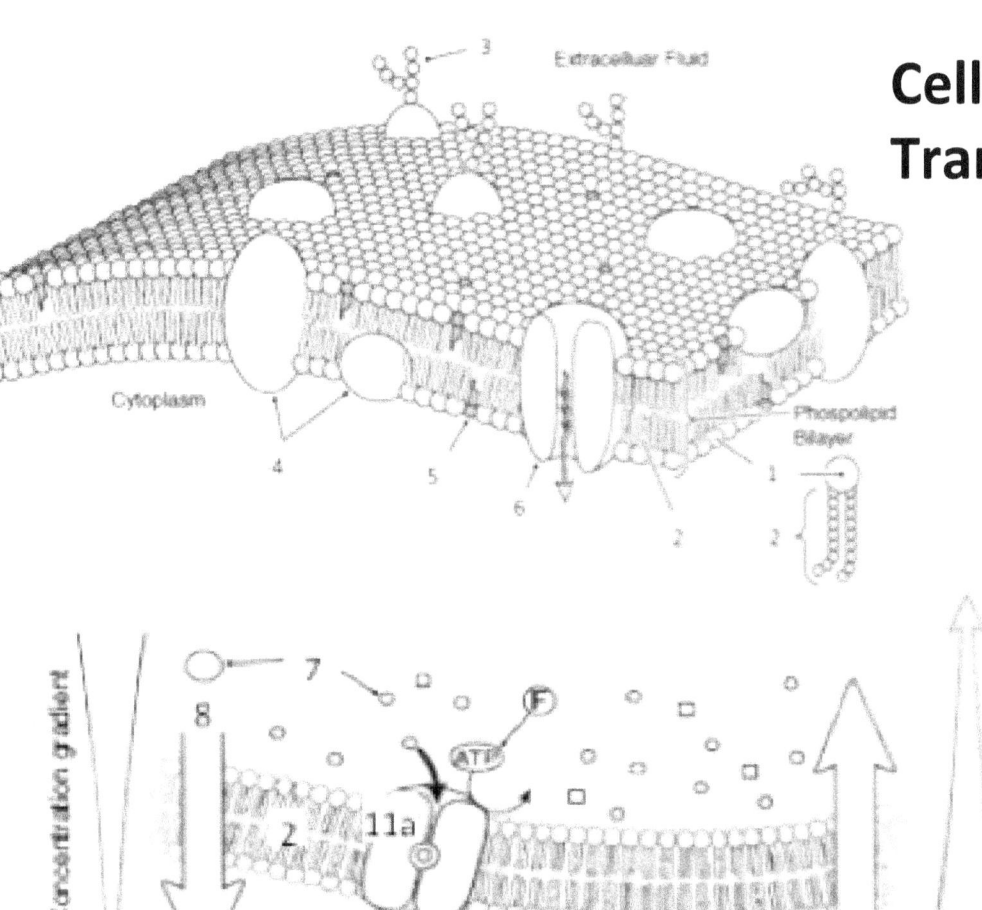

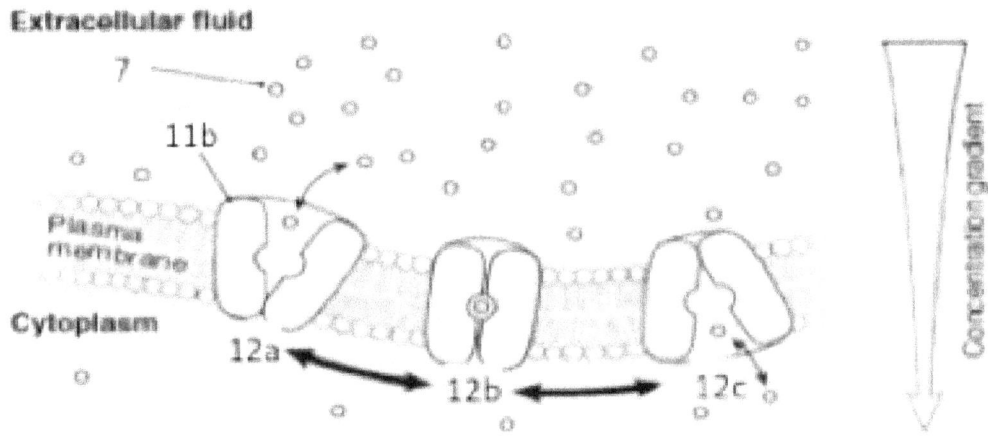

7 Solute Molecule
8 Diffusion
9 Water Molecule
10 Osmosis

11a Active Transport
11b Facilitated Transport
12a Extracellular "Open"
12b Transportation
12c Intracellular "Open"

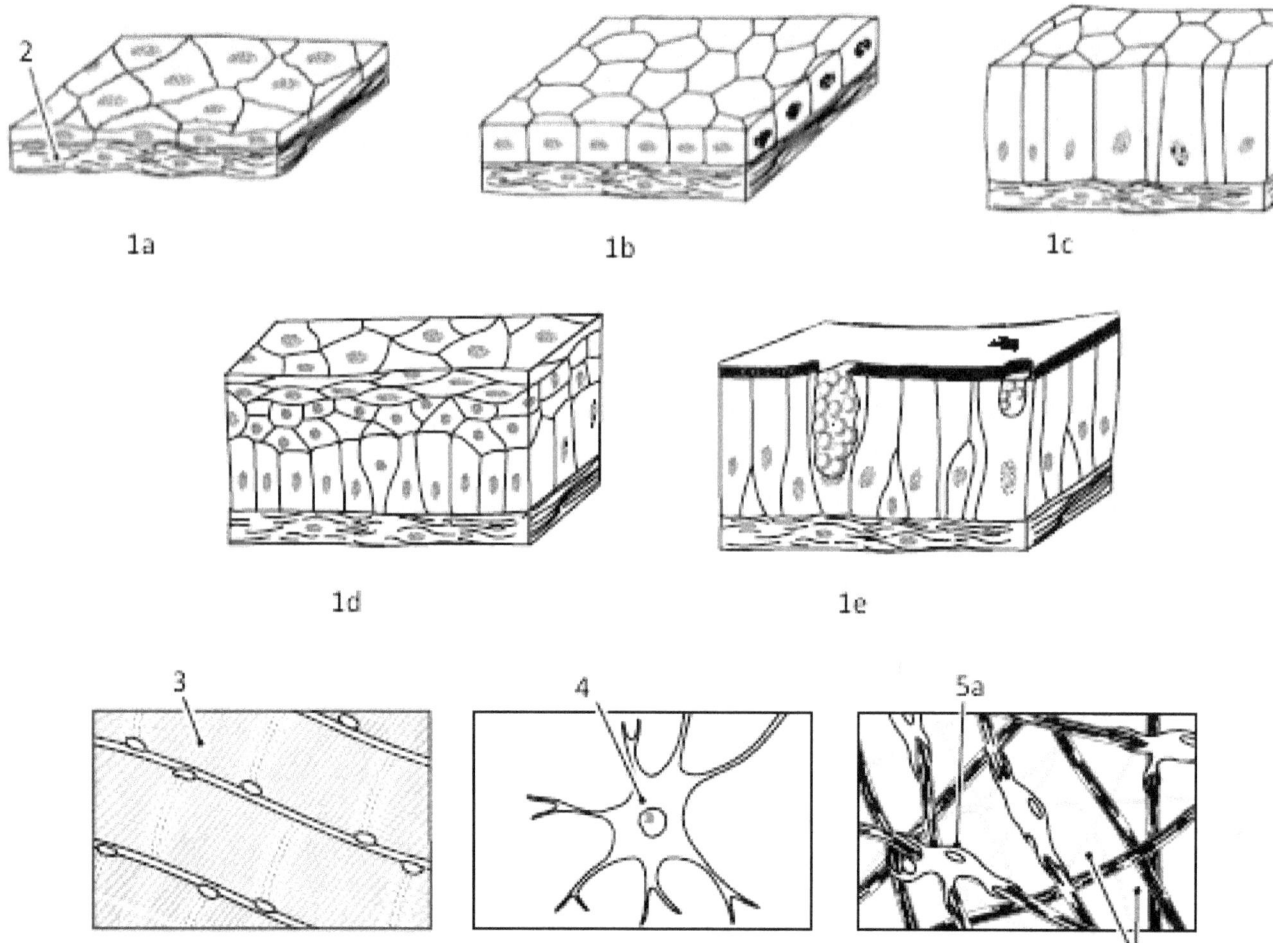

Cells and tissues that make the organs of the body

1a=Squamous Epithelial
1b=Cubodial Epithelial
1c=Columnar Epithelial
1d=Transitional Epithelial
1e=Pseudostratified Epithelial
2=Basal Membrane
3=Striated Muscle
4=Neuron
5a=Fibroblast
5b=Protein Matrix

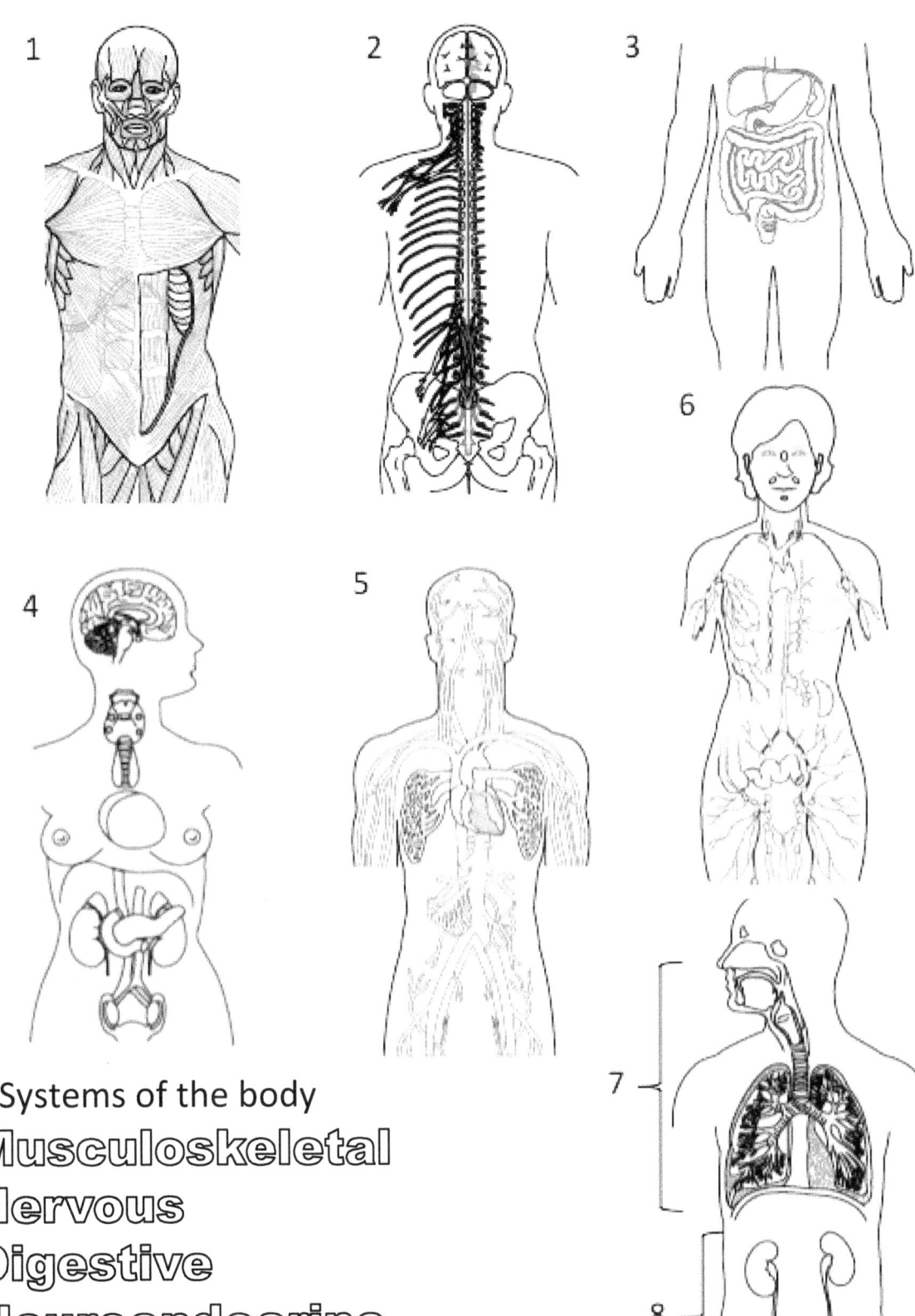

Organ Systems of the body
1. Musculoskeletal
2. Nervous
3. Digestive
4. Neuroendocrine
5. Cardiovascular
6. Immune/Lymphatic
7. Respiratory
8. Excretory

Integument System

Skin and structures of the skin

Overview and Function

The integument (skin) is the largest organ of the body contains many layers of cells and tissues that cover and protect the body from the external environment. The outer most layer is the EPIDERMIS, made of special stratified epithelial cells called KERATINOCYTES. The epidermis acts as a protection layer to loss of water and with MELANOCYTES against ultra-violet light. Then there is the DERMIS that is made of connective tissues and is home to blood vessels, the accessory organs of the integument, such as the SWEAT AND SEBACEOUS GLANDS, the SENSORY STRUCTURES and the beginning of the HAIR AND NAILS that we see on the outer surface. Deep to the dermis is the HYPODERMIS which is connective tissues and adipose (fat) tissue that acts to bind the integument to the the rest of the body and insulates the body.

The sensory organs give use the sense of touch, pressure and temperature. Hair and nail protect the skin from damage, allow us to grip and are involved with grooming and feeding behaviors. Both hair and nails grow from the keratinocytes of the basal membrane within the nail root or hair follicle, the difference is where hair spirals during the growth while nail grow into a flat plate. The color of the skin and hair are dependent on special pigment, melanin, that is made by melanocytes.

Overall the integument serves the following functions for the body:
- Protects the body's internal tissues and organs
- Protects against invasion by infectious organisms
- Protects the body from dehydration
- Protects the body against abrupt changes in temperature (thermoregulation)
- Helps dispose of waste materials
- Acts as a receptor for touch, pressure, pain, heat and cold
- Stores water, fat, and vitamin D.

Integument: Structures

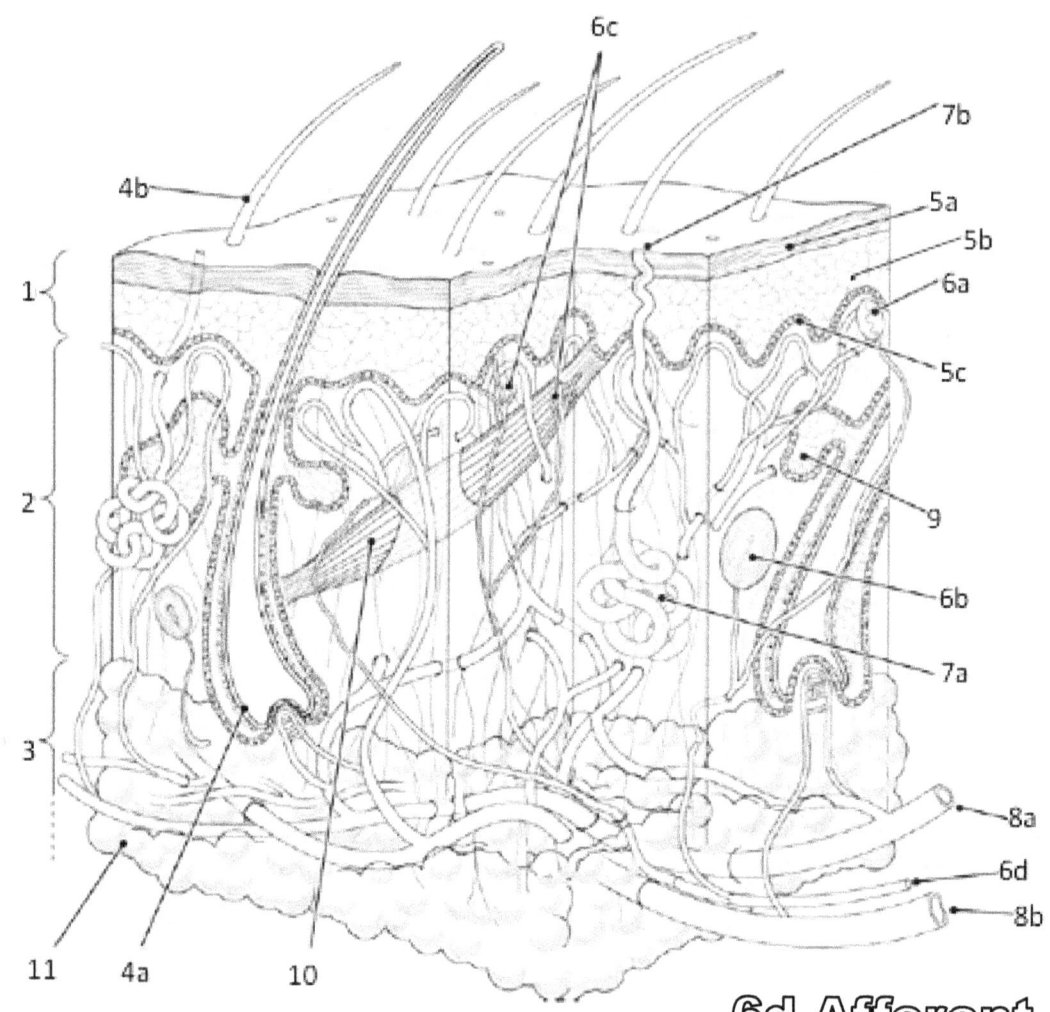

1 Epidermis
2 Dermis
3 Hypodermis (Adipose/Subcutaneous)
4a Hair Follicle
4b Hair Proper
5a Stratified Keratinocytes
5b Stratum Basale
5c Dermal Papilla/Ridges
6a Meissner's Corpuscles
6b Pacinian Corpuscles
6c Merkle's Discs, Nociceptors, Thermoreceptors
6d Afferent Nerve
7a Coiled Gland of Sudoriferous (Sweat) Gland
7b Duct of Sudoriferous (Sweat) Gland
8a Dermal Arterioles
8b Dermal Venuoles
9 Sebaceous Gland
10 Erector Pili Muscle
11 Adipose Tissue

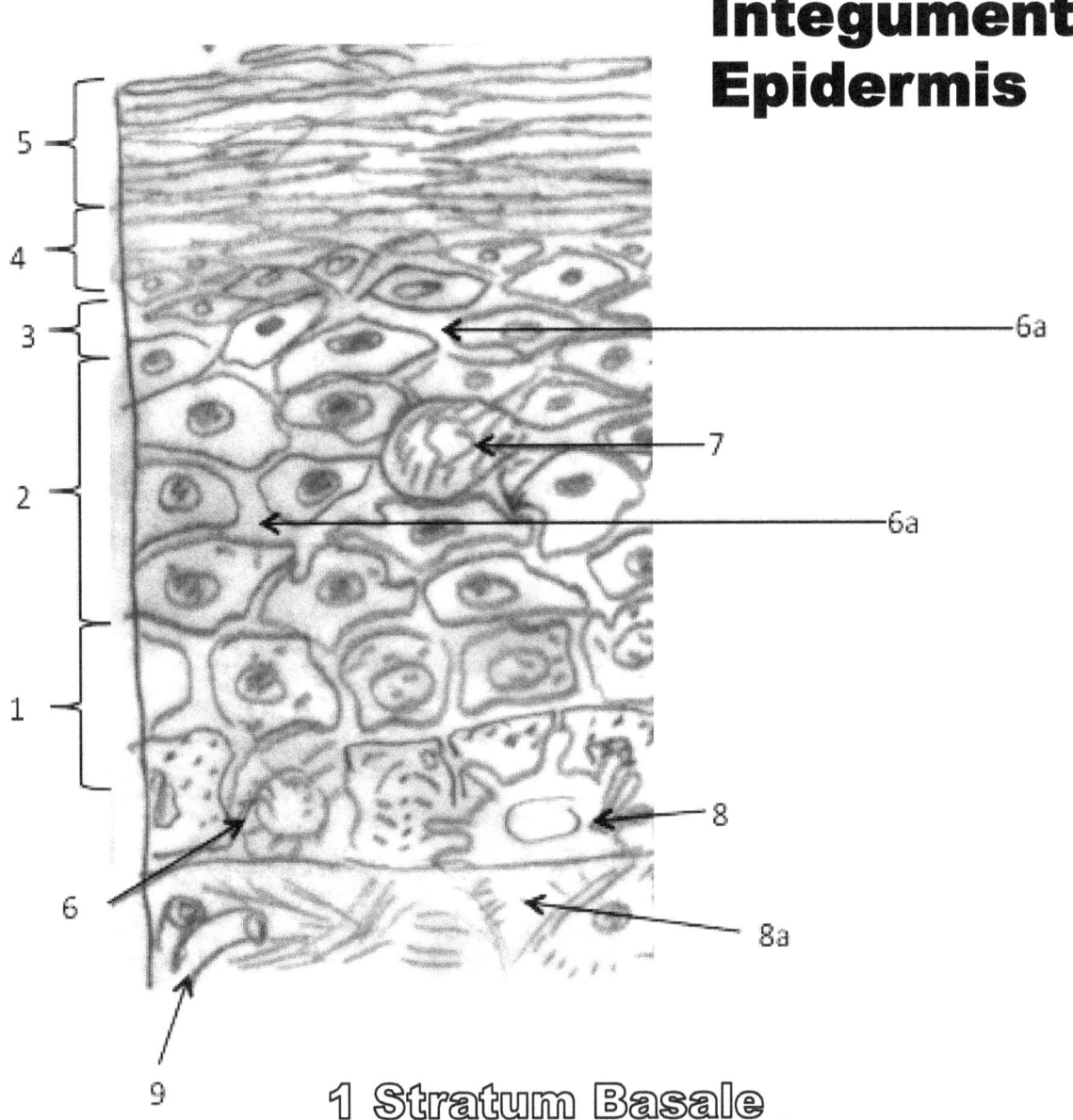

Integument: Epidermis

1 Stratum Basale
2 Stratum Spinosum
3 Stratum Granulosum
4 Stratum Lucidum
5 Stratum Corneum
6 Melanocyte
6a Pseudopodia of Melanocyte
7 Langerhans Cell
8 Merkle's Discs
8a Neuron of Merkle's Discs
9 Capillary

Integument: Sensory Structures

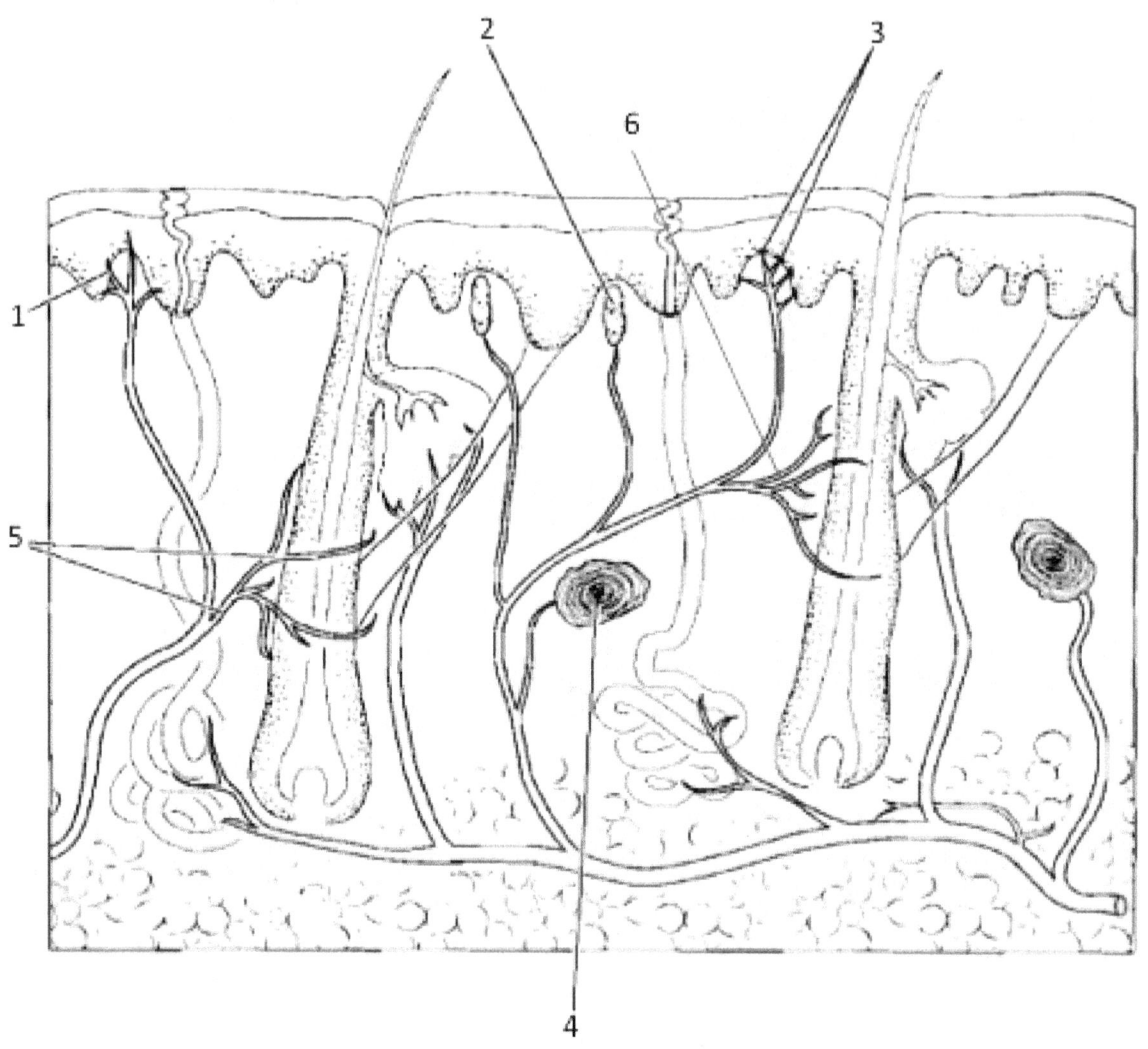

1 Nociceptors & Thermoreceptors
2 Meissner's Corpuscles
3 Merkle's Discs
4 Pacinian Corpuscles
5 Root Hair Plexus
6 Ruffini Endings

Integument: Hair & Nail Structures

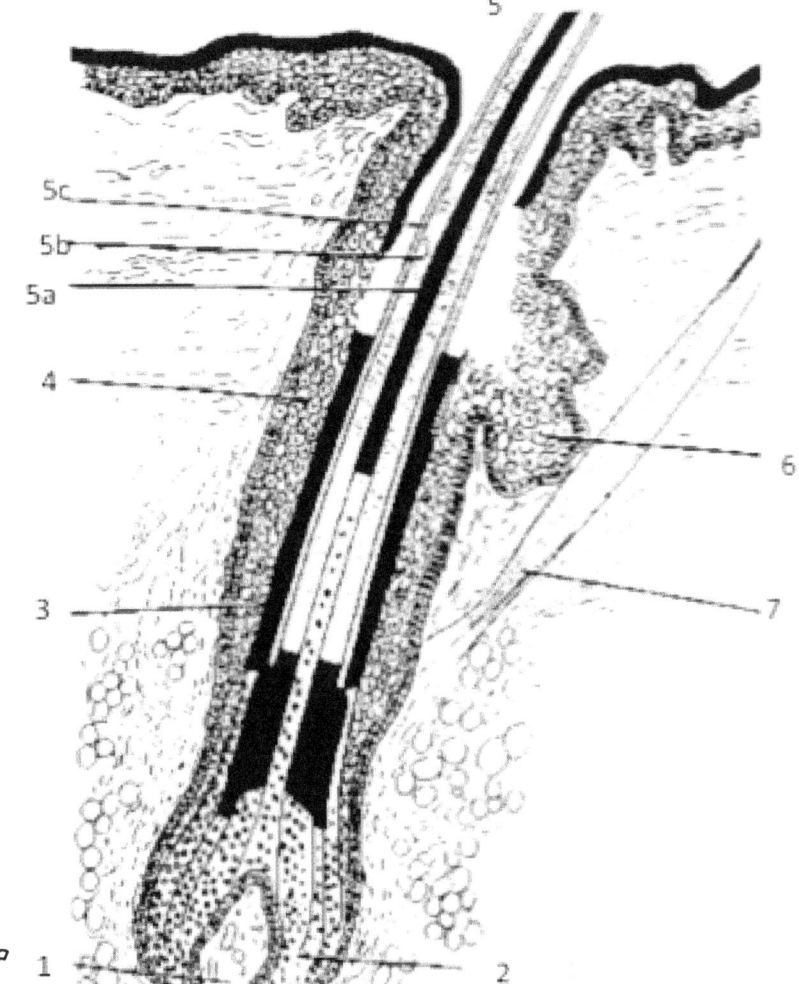

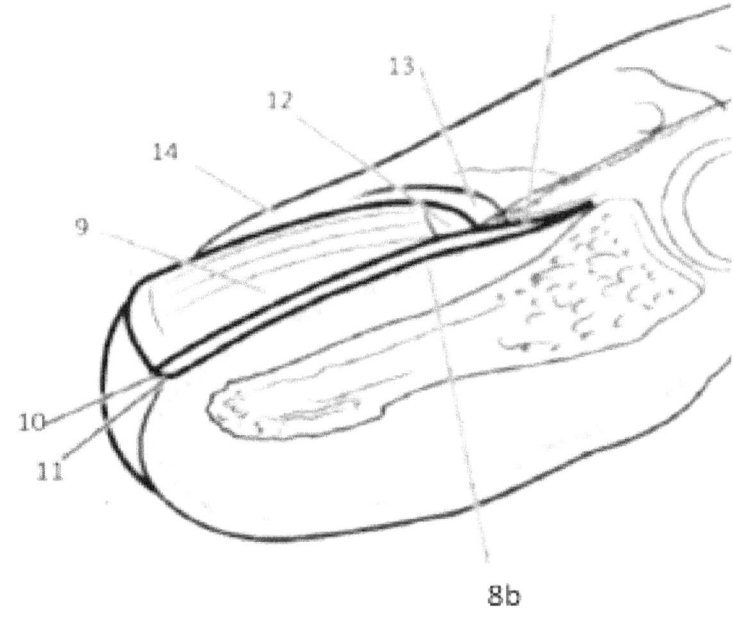

1 Hair Bulb
2 Hair Follicle
3 Hair Root
4 Sheath
5 Hair
5a Medulla of Hair
5b Cortex Of Hair
5c Cuticle of Hair
6 Sebaceous Gland
7 Erector Pili
8a Nail Root
8b Nail Bed
9 Nail Plate
10 Free Edge
11 Hyponychium
12 Lunula
13 Eponychium
14 Nail Fold (Cuticle)

Integument: Glands of the Body

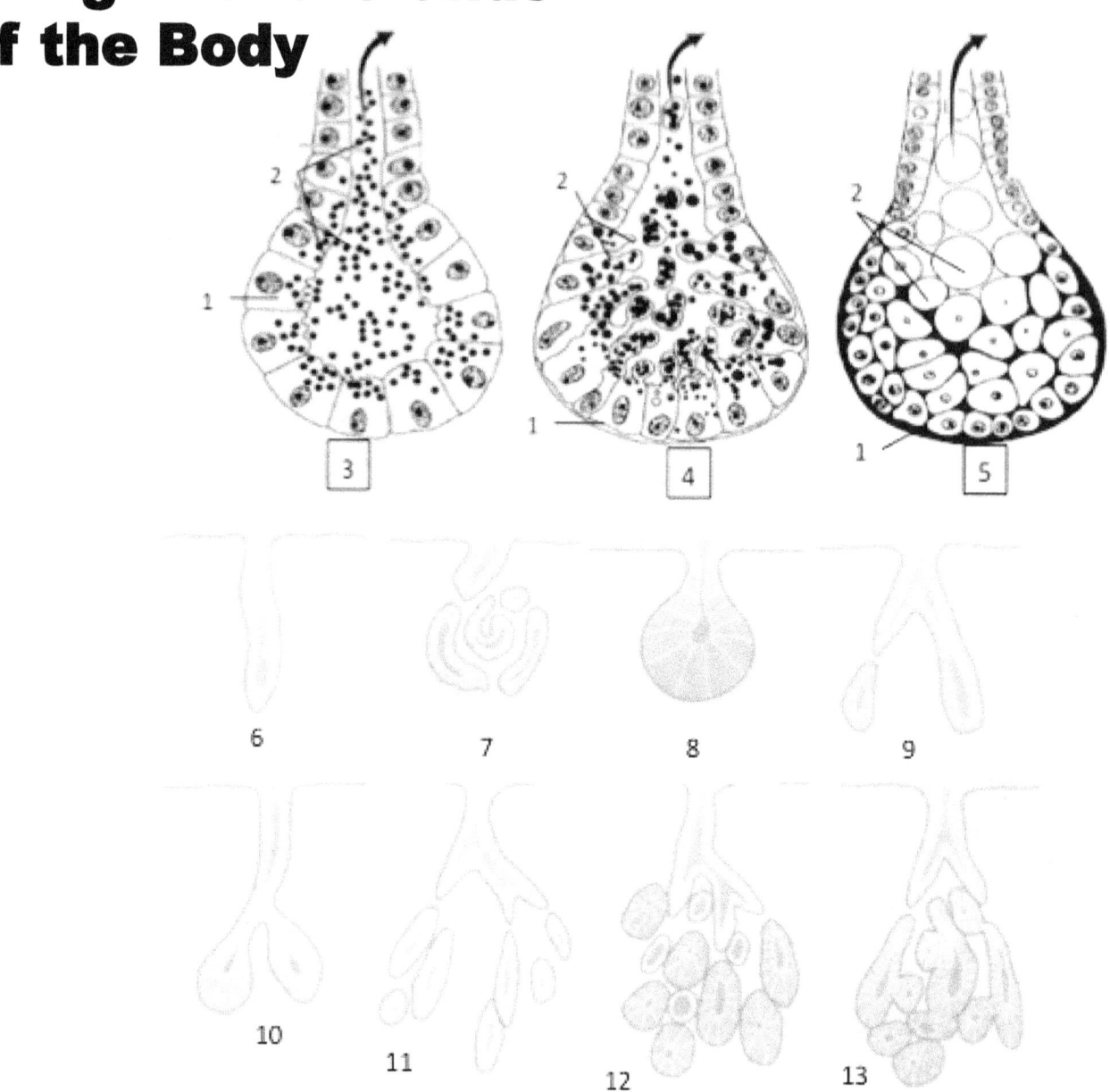

1 Secretory Cell
2 Secretion
3 Merocrine
4 Apocrine
5 Halocrine
6 Simple Tubular
7 Simple Coiled
8 Simple Acinar
9 Simple Branched Tubular
10 Simple Branched Acinar
11 Compound Branched Tubular
12 Compound Branched Acinar
13 Compound Branched Tubuloacinar

Musculoskeletal System

Bones, Skeletal Muscles
and Joints of the body

Skeletal Overview

The human skeleton can be divided into AXIAL and APPENDICULAR skeleton. In describing the bones that make these structures there are four basic classifications of bones based on the shape of the bone: FLAT, such as the skull, LONG, such as the femur or humerus, SHORT, such as the fingers, and IRREGULAR, such as the vertebrae. There are also various landmarks and structures that give distinct regionalization to the bone. Some of these landmarks provide the platform for movement, while others allow for nerves and blood vessels to pass through the bone, and others are the attachment points for tendons of the skeletal muscles that move the bones the body. Each bone within the body (all 200+ bones) will have some combination of the various landmarks based on the function, and tissues attached to, each of the bones.

Movement of bones within the skeleton occurs at ARTICULATIONS (JOINTS) of the skeleton. These can either provided for no relative movement (SYNARTHOSIS), as it is simply where two bones meet, some movement (AMPHIARTHROSIS) or full range of motion (DIARHTOSIS) between bones that are meeting at the joint. Diarthrosis are also called SYNOVIAL and are classified as HINGE, GLIDING, ELLIPSOID, SADDLE, PIVOT, and BALL-AND-SOCKET based on the shape of the ends of the bones that are meeting.

Musculoskeletal: Bone Structure 1 - Landmarks & Anatomy of the bone

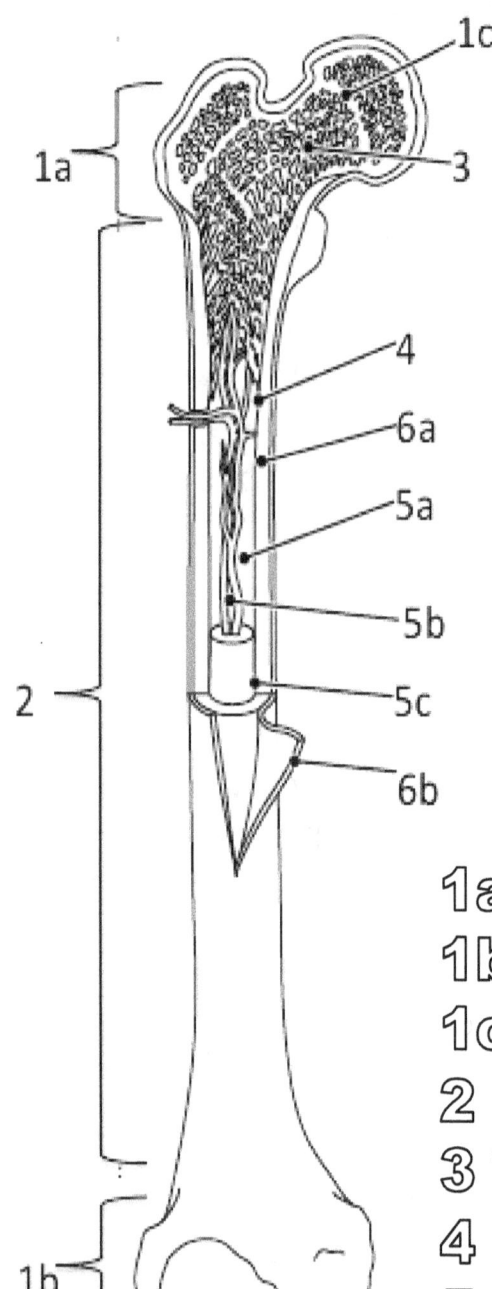

1a Proximal Epiphyseal
1b Distal Epiphyseal
1c Epiphyseal Line
2 Diaphyseal (Body of Bone)
3 Trabecular (Spongy) Bone
4 Cortical (Compact) Bone
5a Medullar Cavity
5b Artery/Vein
5c Yellow Marrow (Adipose)
6a Endosteum (lining inside)
6b Periosteum (covering outside)

Musculoskeletal: Bone Structure 2-Tissue

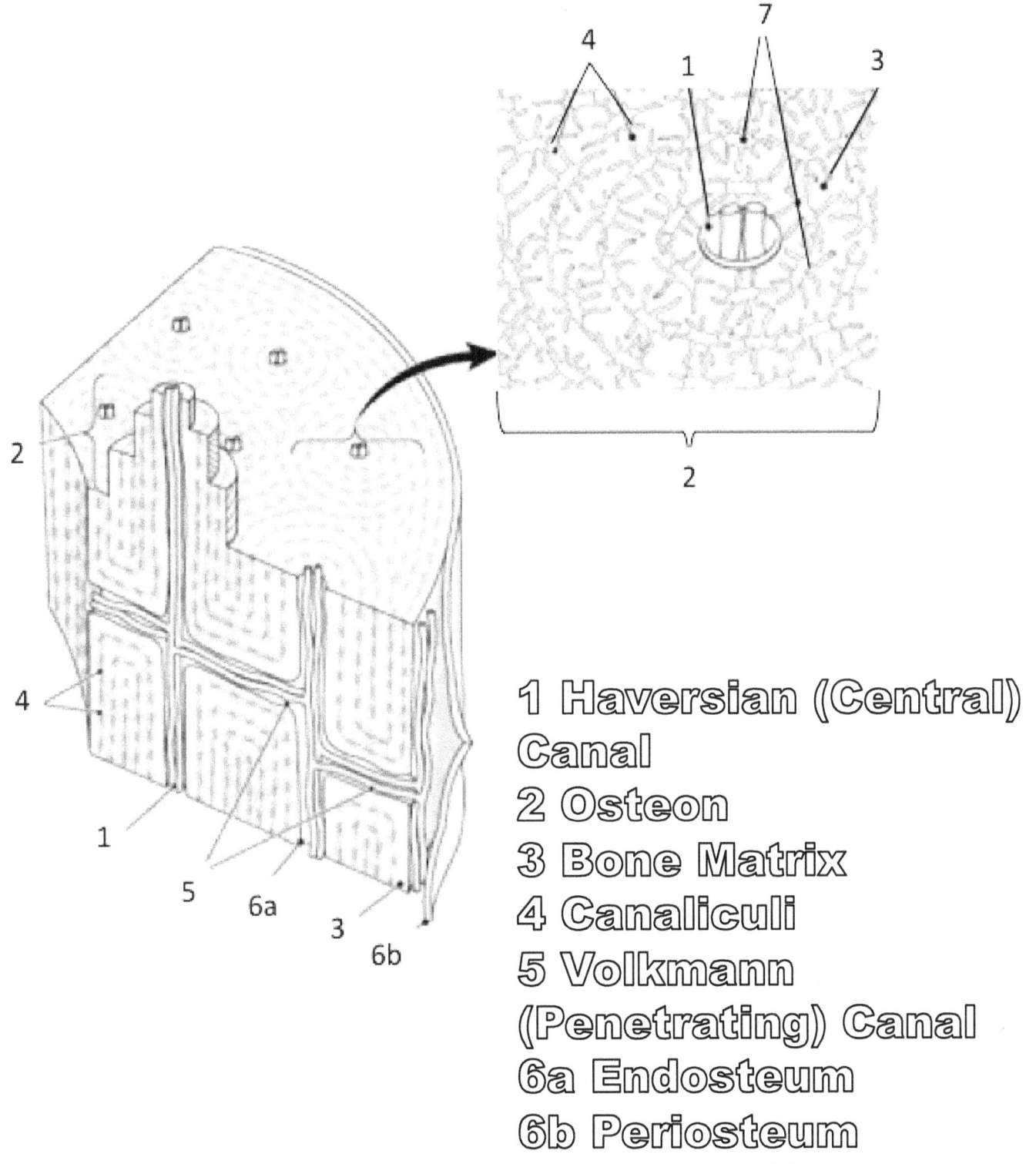

1 Haversian (Central) Canal
2 Osteon
3 Bone Matrix
4 Canaliculi
5 Volkmann (Penetrating) Canal
6a Endosteum
6b Periosteum
7 Lacunae/Osteocyte

Musculoskeletal: Articulations 1- Joint Types

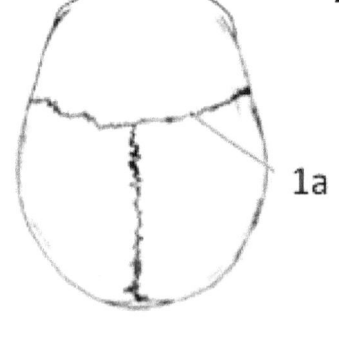
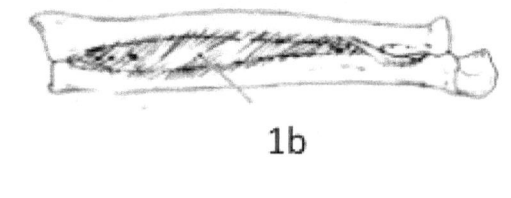
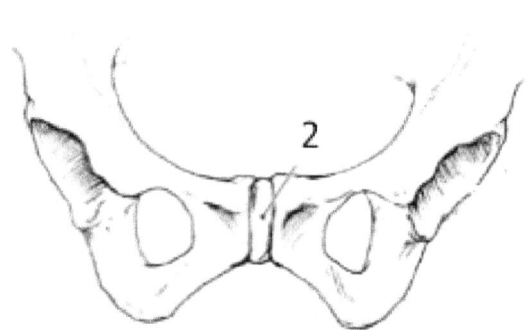
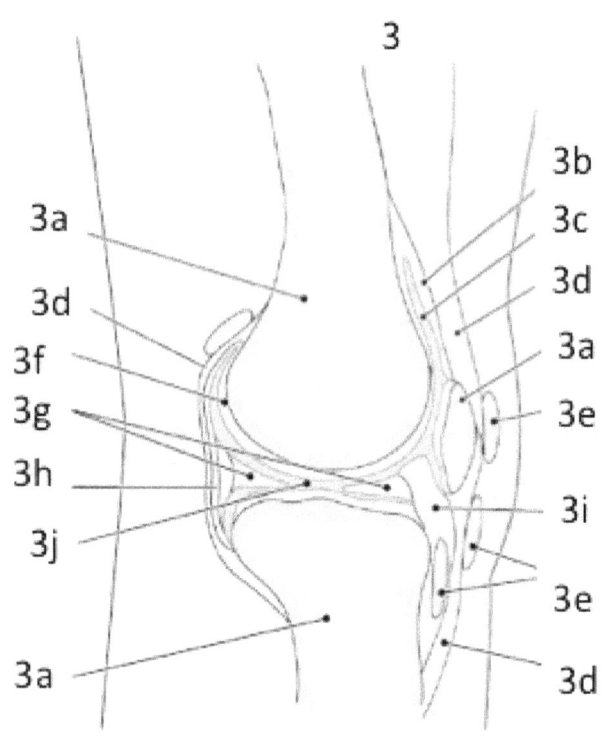

1a Fibrous (Synarthrosis)
1b Fibrous (Amphiarthrosis)
2 Cartilaginous (Amphiarthrosis)
3 Synovial

3a Articulating Bones
3b Anterior Synovial Membrane
3c Synovial Fluid
3d Tendon
3e Bursa Sacs
3f Articulating Cartilage
3g Meniscus
3h Posterior Synovial Membrane
3i Sub-patellar Bursa Sac
3j Cruciate Ligament

Musculoskeletal: Articulations 2-Synovial Joints

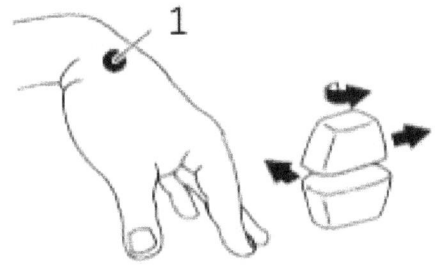
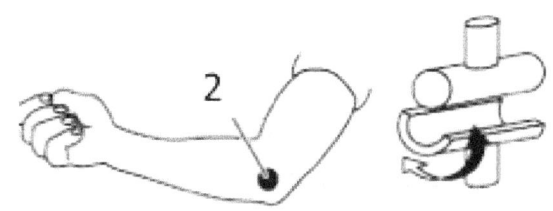
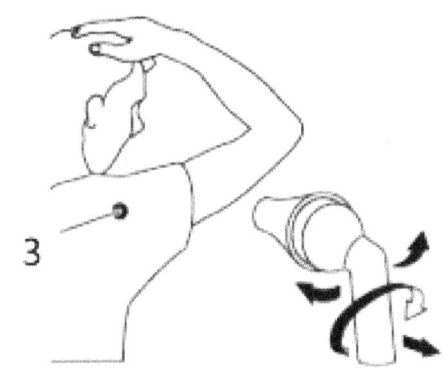
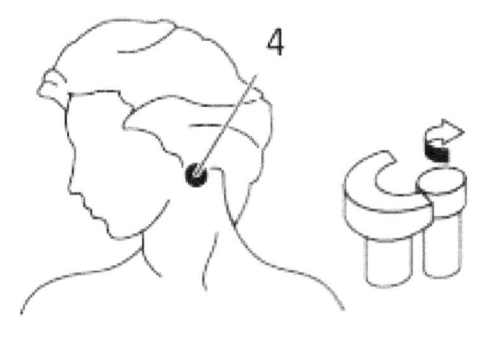
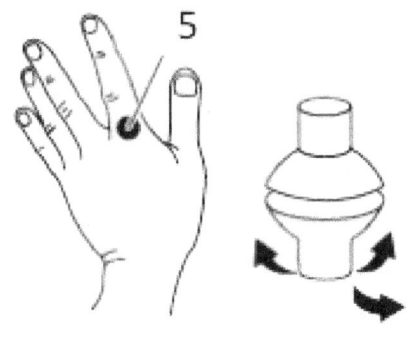
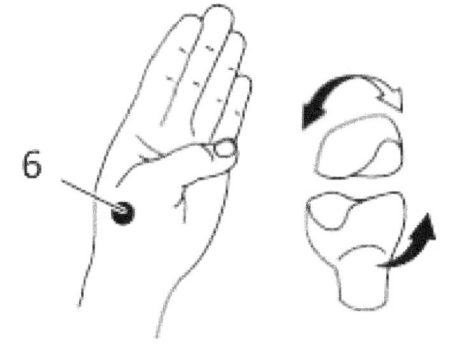

1 Gliding
2 Hinge
3 Ball-and-Socket
4 Pivot
5 Ellipsoid
6 Saddle

Boney Landmarks

In describing bones, we typically note various landmarks and structures that give distinct regionalization to the bone. Some of these landmarks provide the platform for movement, while others allow for nerves and blood vessels to pass through the bone, and others are the attachment points for tendons of the skeletal muscles that move the bones the body. Each bone within the body (all 200+ bones) will have some combination of the various landmarks that are being described here and the diagrams of the major bones will describe in detail each of the landmarks for each of the bones.

HEAD: typically describes the proximal articulating surface of a long bone
NECK: describes the area of the bone that connects the head of the bone to the remainder of the bone
SHAFT: the length of the bone from the head and neck to the condyles
TROCHANTER/TUBEROSITY: tendon attachment point, the name dictates the size of the attachment point
EPICONDYLE: broad area of the bone superior to the condyle of the bone, site of many collateral ligament attachment points of the synovial joints
CONDYLE: the distal articulating surface of a bone
SPINE: a elongated projection of bone that provides a surface for attachment of tendon and ligament
FOSSA: an indentation, or flattened surface, that provides an area for convergent muscle to originate
PROCESS: generic term for any boney projection
FORAMEN: an opening in the bone that provides an area for blood vessels, nerves, and muscles to pass through the bone

Musculoskeletal: Skeleton 1-Overview

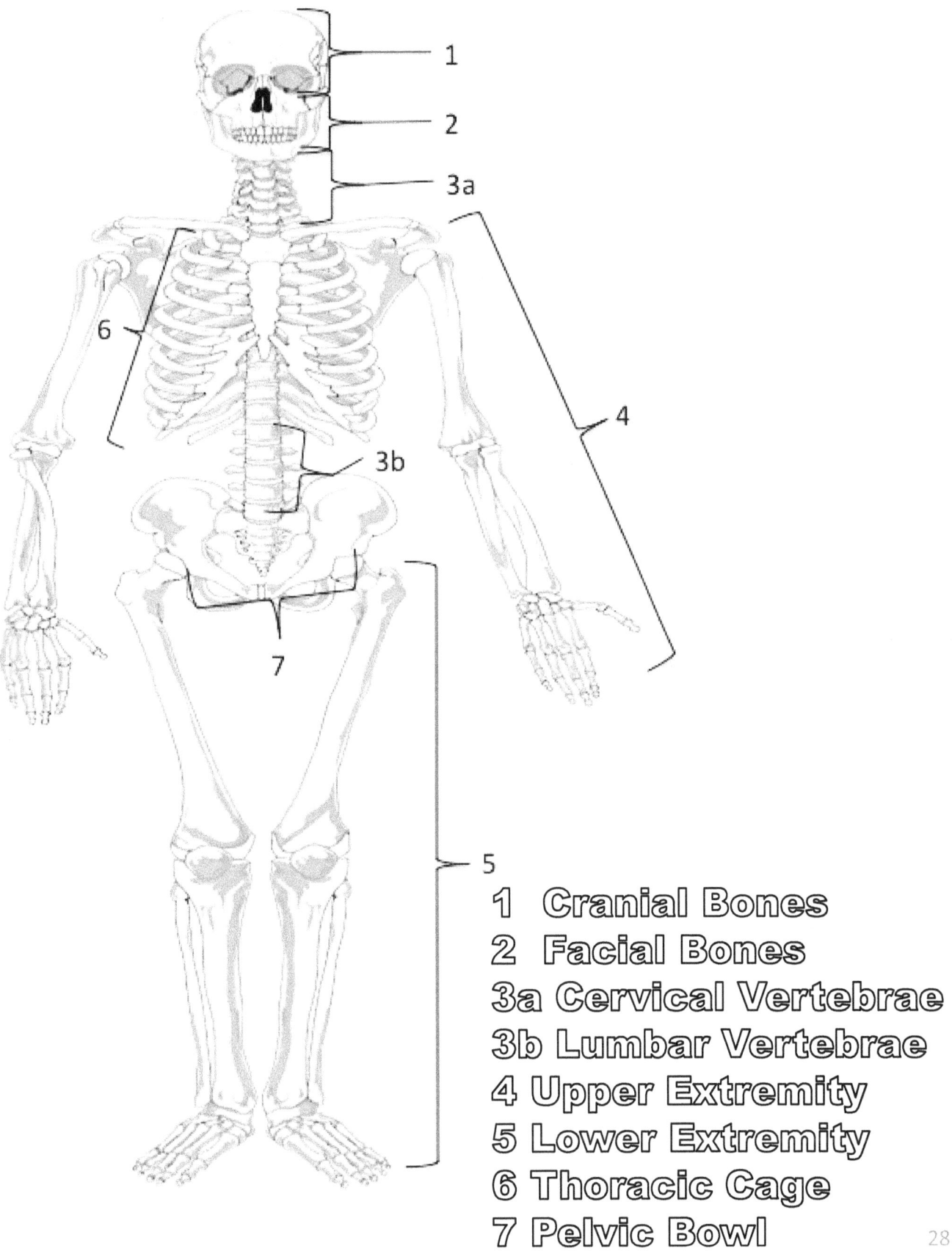

1 Cranial Bones
2 Facial Bones
3a Cervical Vertebrae
3b Lumbar Vertebrae
4 Upper Extremity
5 Lower Extremity
6 Thoracic Cage
7 Pelvic Bowl

Musculoskeletal: Skeleton
2-Cranial & Facial Bones

1 Frontal
2 Nasal
2a Nasal Conchae
3 Parietal
4 Temporal
4a Styloid Process
4b Mastoid Process
4c Zygomatic Process
4d External Auditory Meatus
5 Sphenoid
6 Zygomatic
7 Occipital
8 Maxilla
8a Maxillary Teeth
9 Mandible
9a Mental Foramen
9b Mandibular Teeth
9c Ramus of the Mandible
10 Ethmoid
11a Palatine
11b Vomer
12 Hyoid
13a Coronal Suture
13b Sagittal Suture
13c Squamous Suture
13d Lamboid Suture

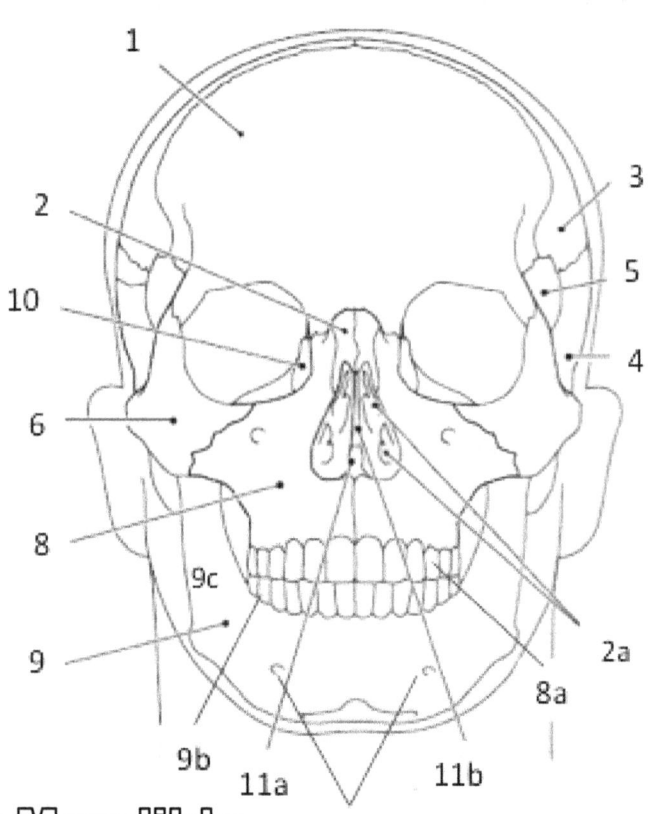
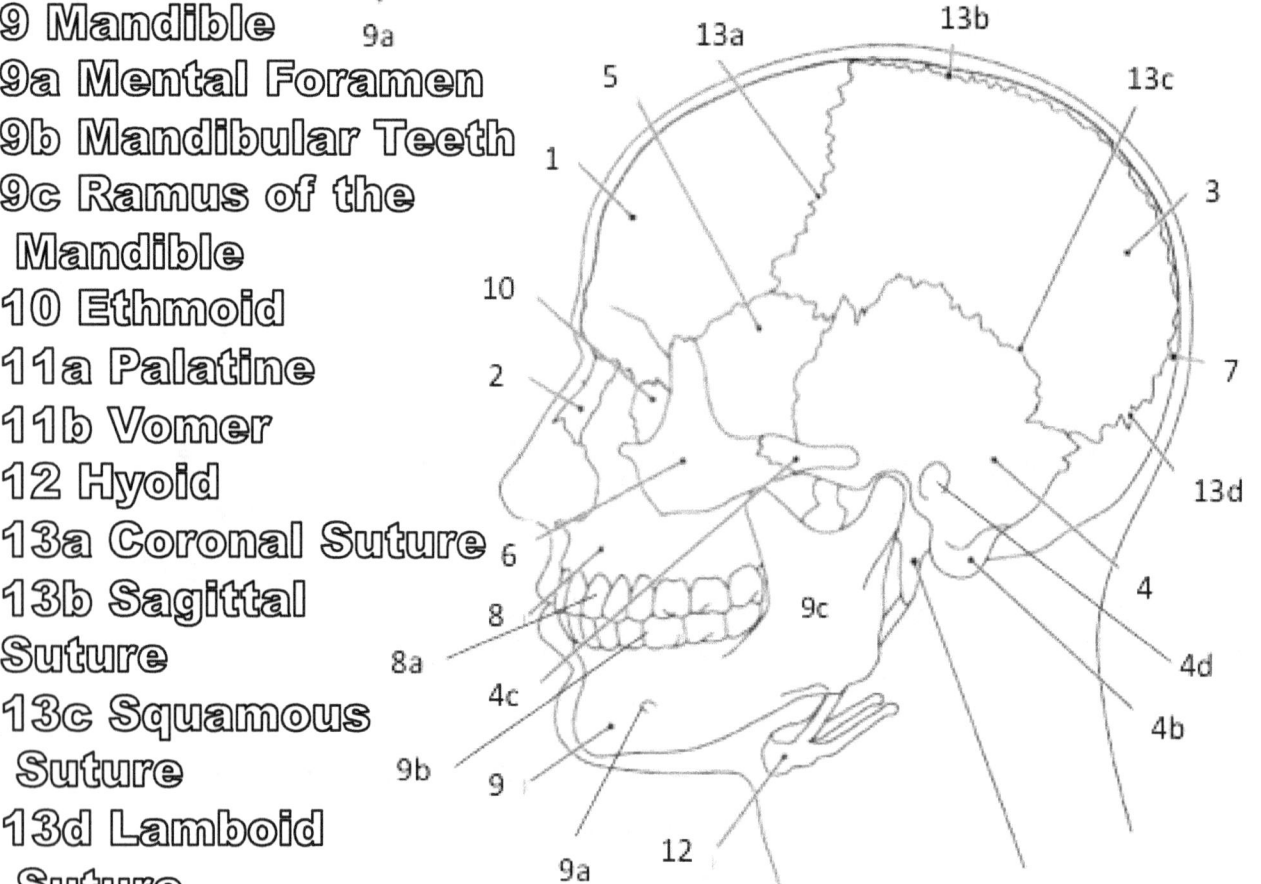

Musculoskeletal: Skeleton
3-Vertebrae & Curvatures

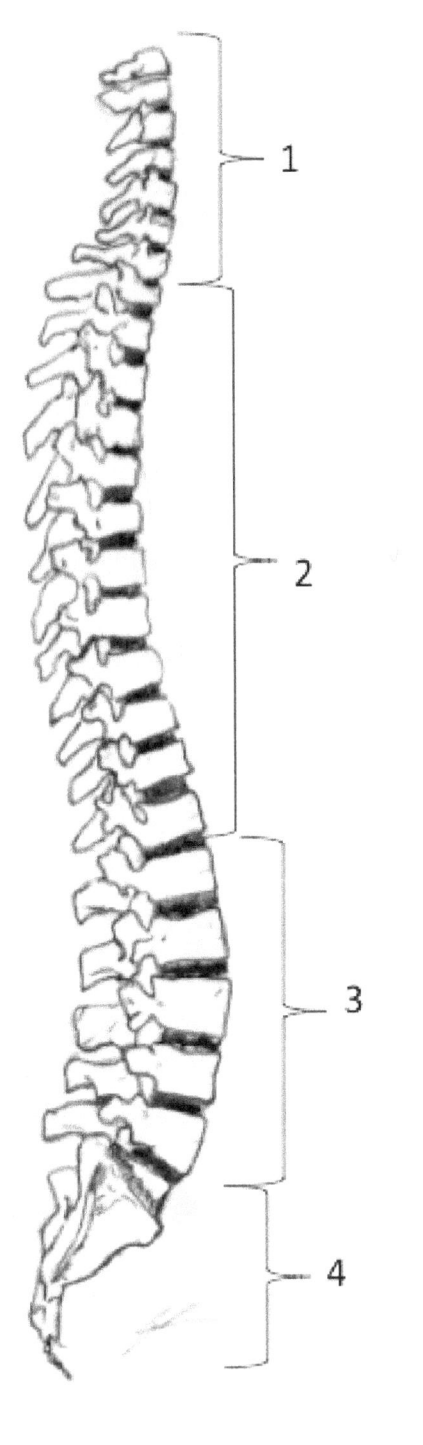
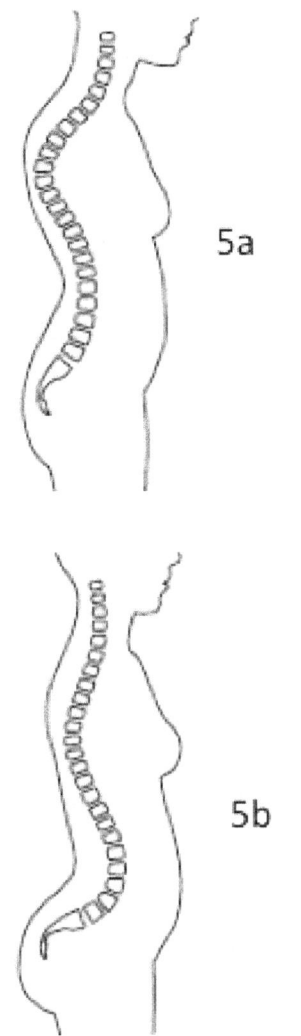
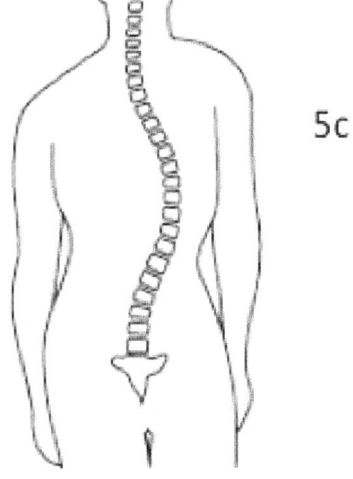

1 Cervical Vertebrae
2 Thoracic Vertebrae
3 Lumbar Vertebrae
4 Sacral/Coccygeal

5a Hyperkyphosis
5b Hyperlorodosis
5c Scoliosis

Musculoskeletal: Skeleton 4-Vertebrae Identification & Landmarks

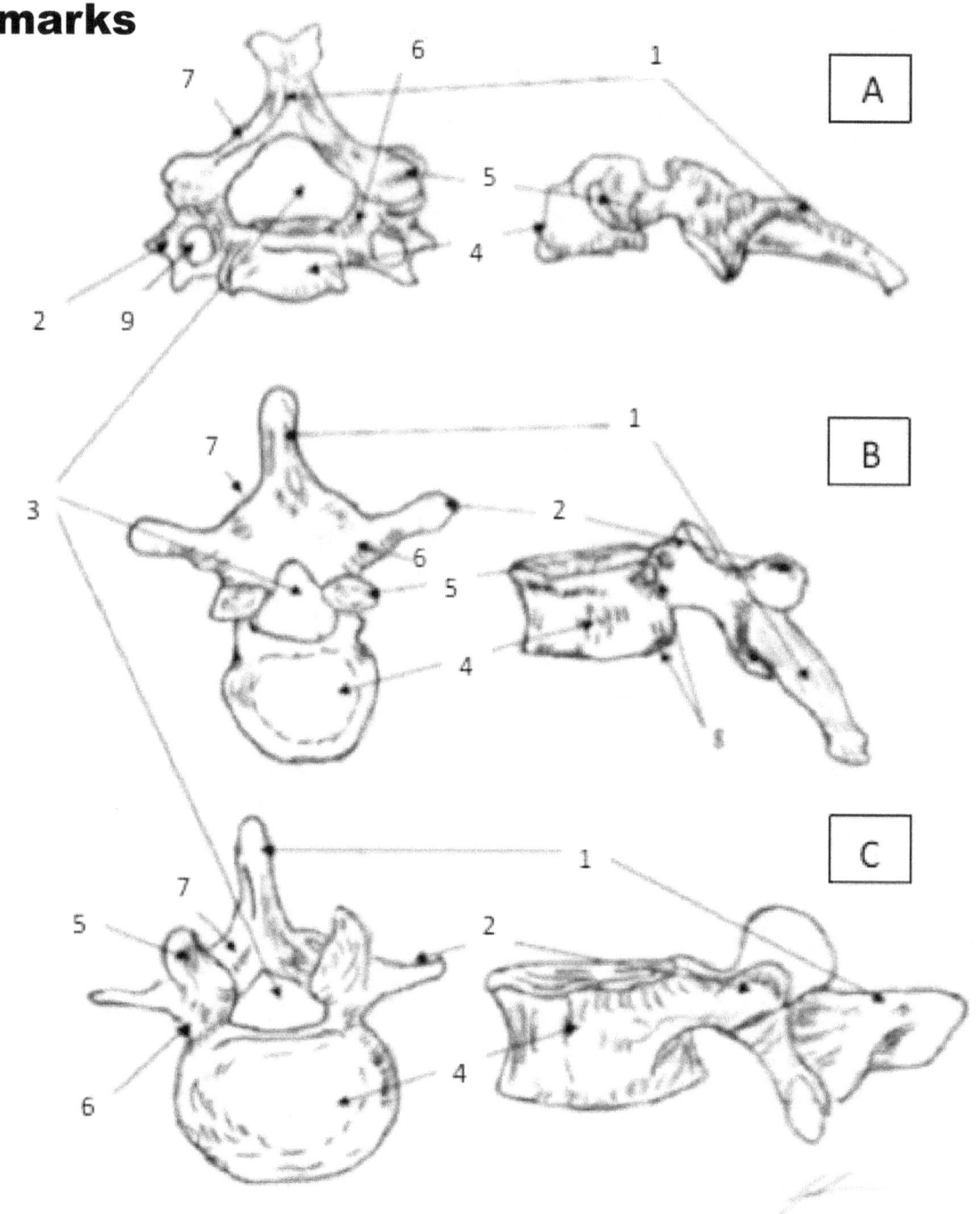

A Cervical
B Thoracic
C Lumbar
1 Spinal Process
2 Transverse Process
3 Vertebral Foramen
4 Body of Vertebrae
5 Articulating Facet of Vertebrae
6 Pedicle
7 Lamina
8 Costal Facets (Only in Thoracic)
9 Transverse Foramen (Only in Cervical)

Musculoskeletal: Skeleton 5-Thoracic Cage & Rib (Costal) Bone Identification & Landmarks

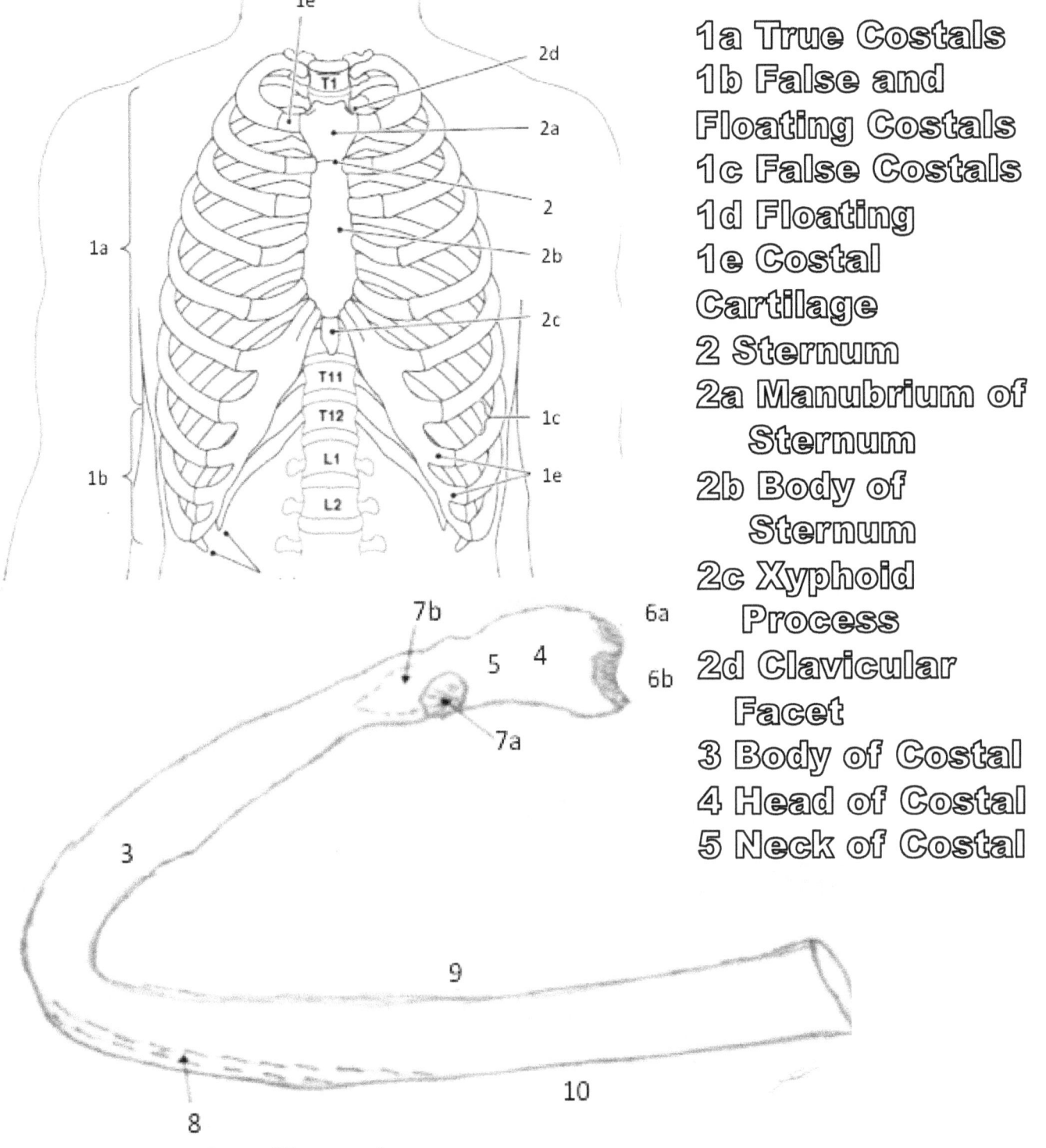

1a True Costals
1b False and Floating Costals
1c False Costals
1d Floating
1e Costal Cartilage
2 Sternum
2a Manubrium of Sternum
2b Body of Sternum
2c Xyphoid Process
2d Clavicular Facet
3 Body of Costal
4 Head of Costal
5 Neck of Costal

6a Superior Facet
6b Inferior Facet
7a Vertebral Facet
7b Vertebral Fossa (Posterior Surface)

8 Costal Groove (Posterior Surface)
9 Round (Superior) Edge
10 Sharp (Inferior) Edge

Musculoskeletal: Skeleton 6-Upper Extremity Bones

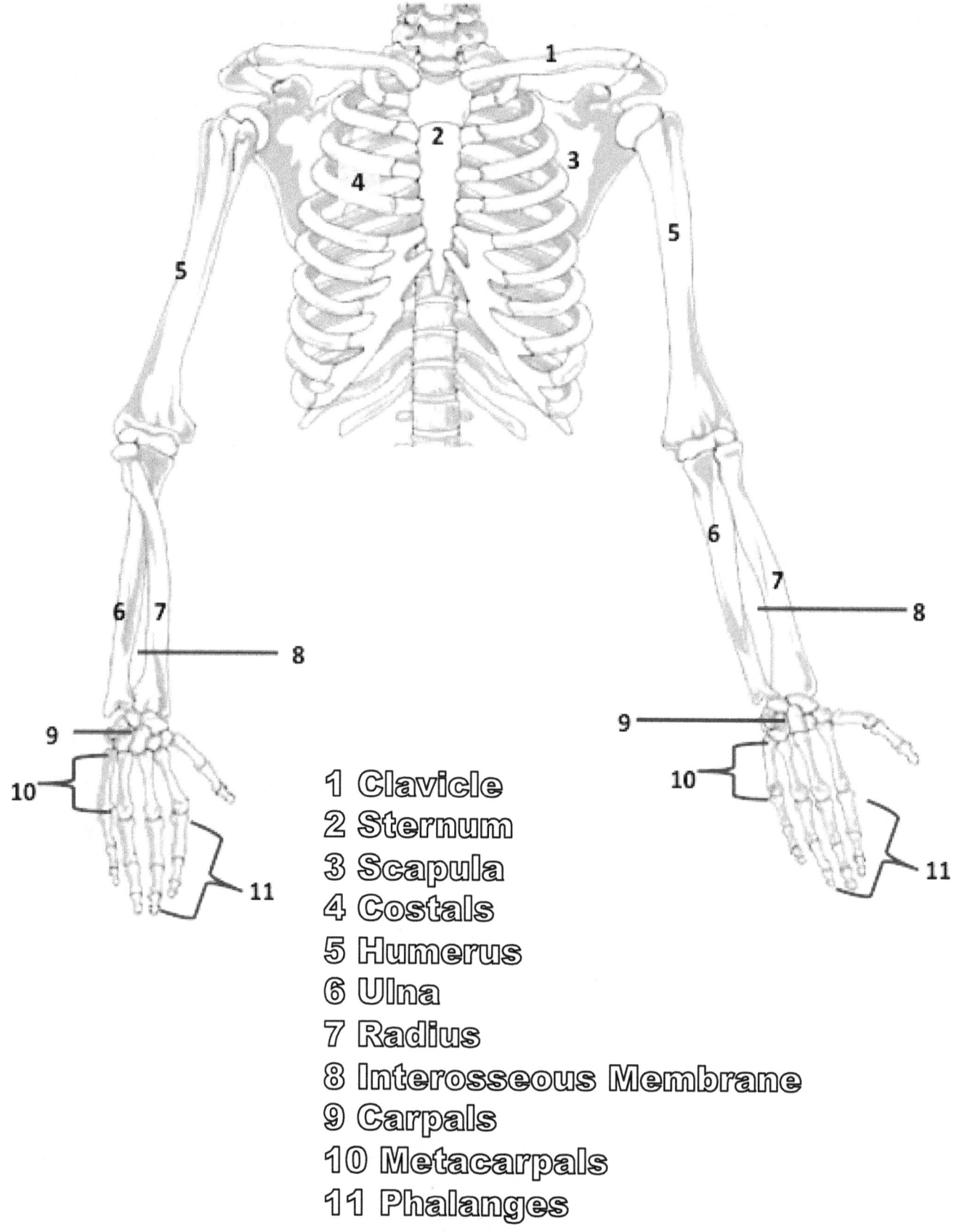

1 Clavicle
2 Sternum
3 Scapula
4 Costals
5 Humerus
6 Ulna
7 Radius
8 Interosseous Membrane
9 Carpals
10 Metacarpals
11 Phalanges

Musculoskeletal: Skeleton
7-Scapula & Clavicle

1 Thoracic Vertebrae
2 Costal #1
3 Manubrium of Sternum
4 Proximal Head of Clavicle
5 Distal Head of Clavicle
6 Neck of Clavicle
7 Spine of Scapula
8 Supraspinatal Fossa
9 Infraspinatal Fossa
10 Glenoid Fossa
11 Lateral Border
12 Coricoid Process
13 Acromion Process
14a Inferior Angle of Scapula
14b Superior Angle of Scapula
15 Medial Border

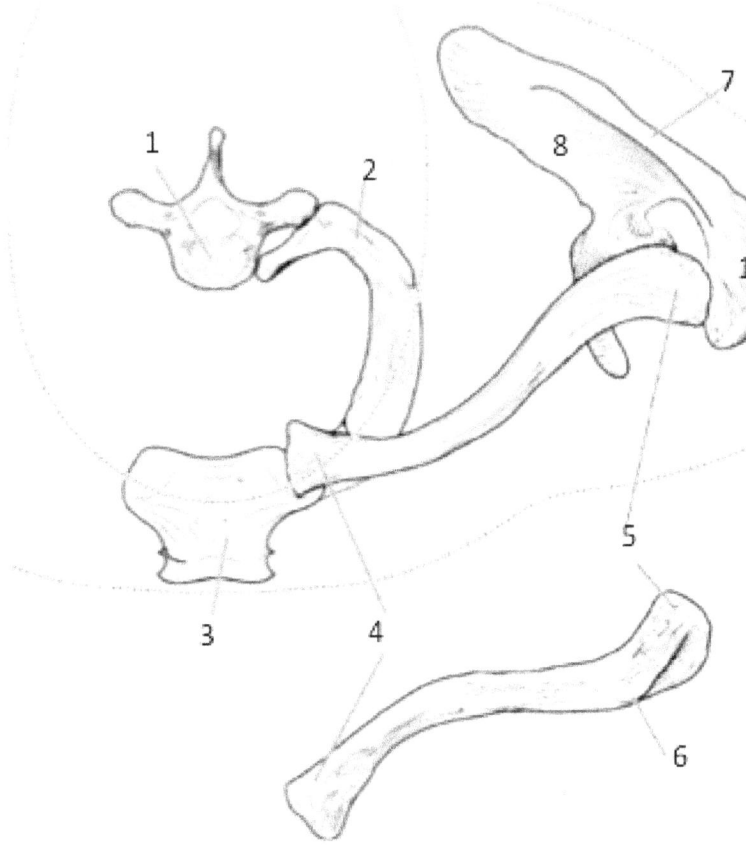
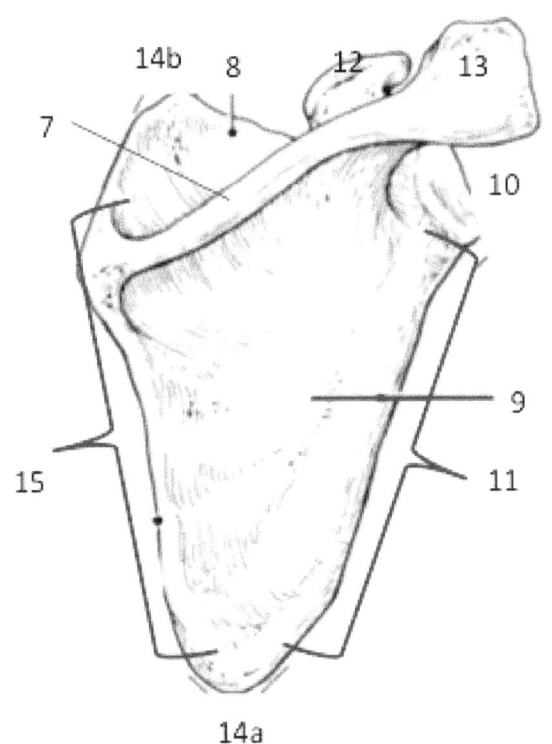
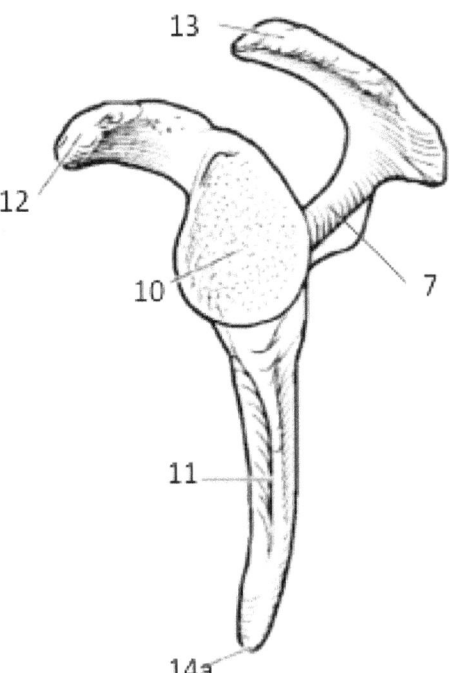

Musculoskeletal: Skeleton 8-Upper Extremity Bones Identification & Landmarks 1 (Humerus)

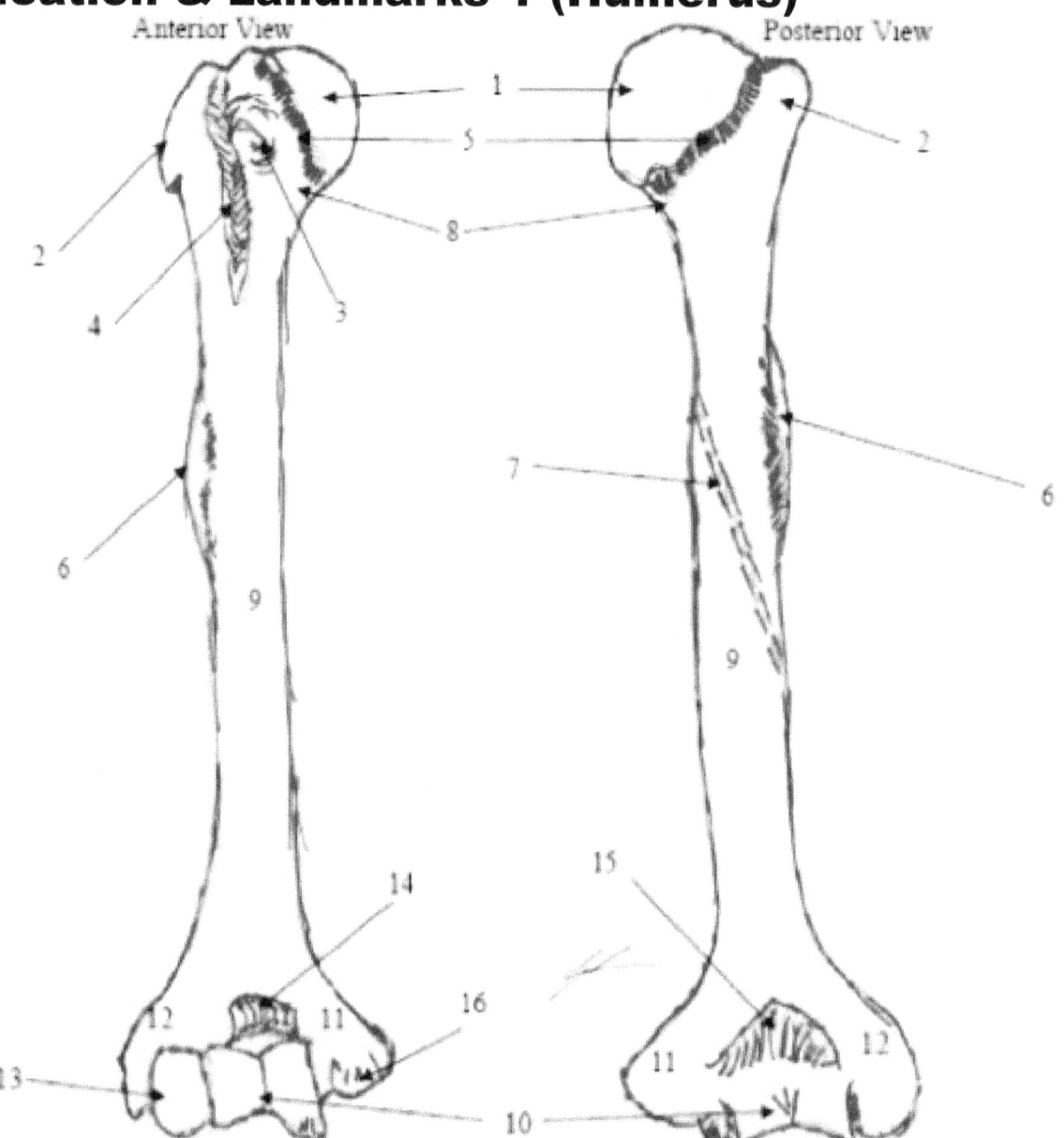

1 Head of Humerus
2 Greater Tubercle
3 Lesser Tubercle
   (only seen anteriorly)
4 Intertubericular Groove
5 Neck of Humerus
6 Deltoid Tuberosity
7 Radial Groove
   (only seen posteriorly)
8 Surgical Neck
9 Body of Humerus
10 Trochlea
11 Medical Epicondyle of Humerus
12 Lateral Epicondyle of Humerus
13 Capitulum
13 Trochlear Notch
15 Olecranon Fossa
16 Radial Fossa

Musculoskeletal: Skeleton 8-Upper Extremity Bones Identification & Landmarks 2 (Radius & Ulna)

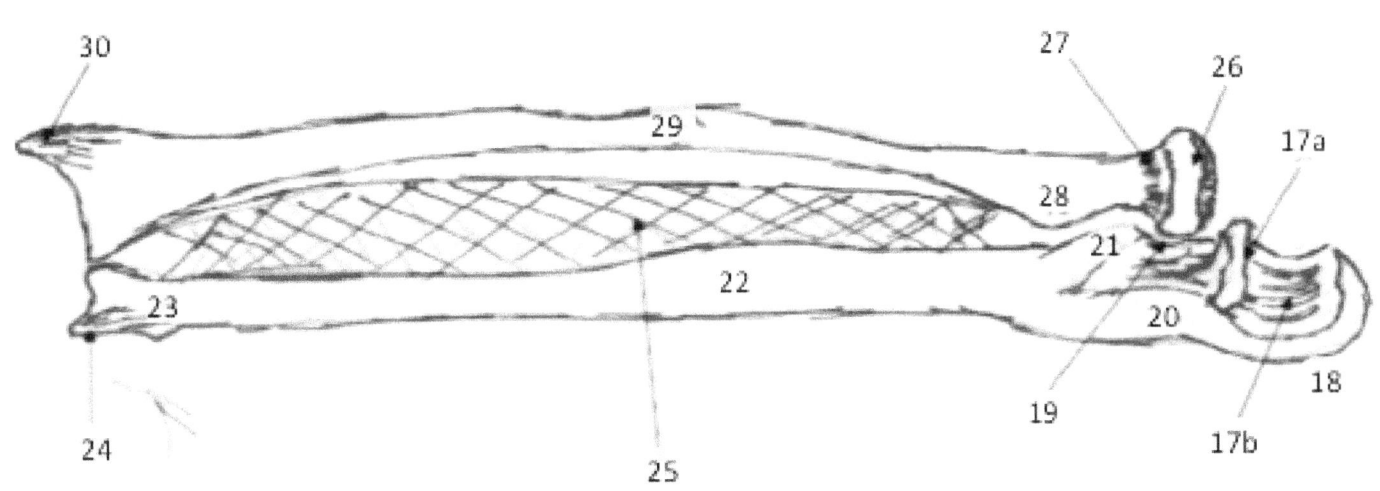

17a Trochlear Notch
17b Coronoid Process
18 Olecranon Process
19 Radial Notch
20 Proximal Head of Ulna
21 Ulnar Tuberosity
22 Body Of Ulna
23 Distal Head of Ulna
24 Ulnar Styloid
25 Interosseous Membrane
26 Head of Radius
27 Neck of Radius
28 Radial Tuberosity
29 Body of Radius
30 Radial Styloid

Musculoskeletal: Skeleton 9- Hand & Wrist (Carpals and Metacarpals) Identification

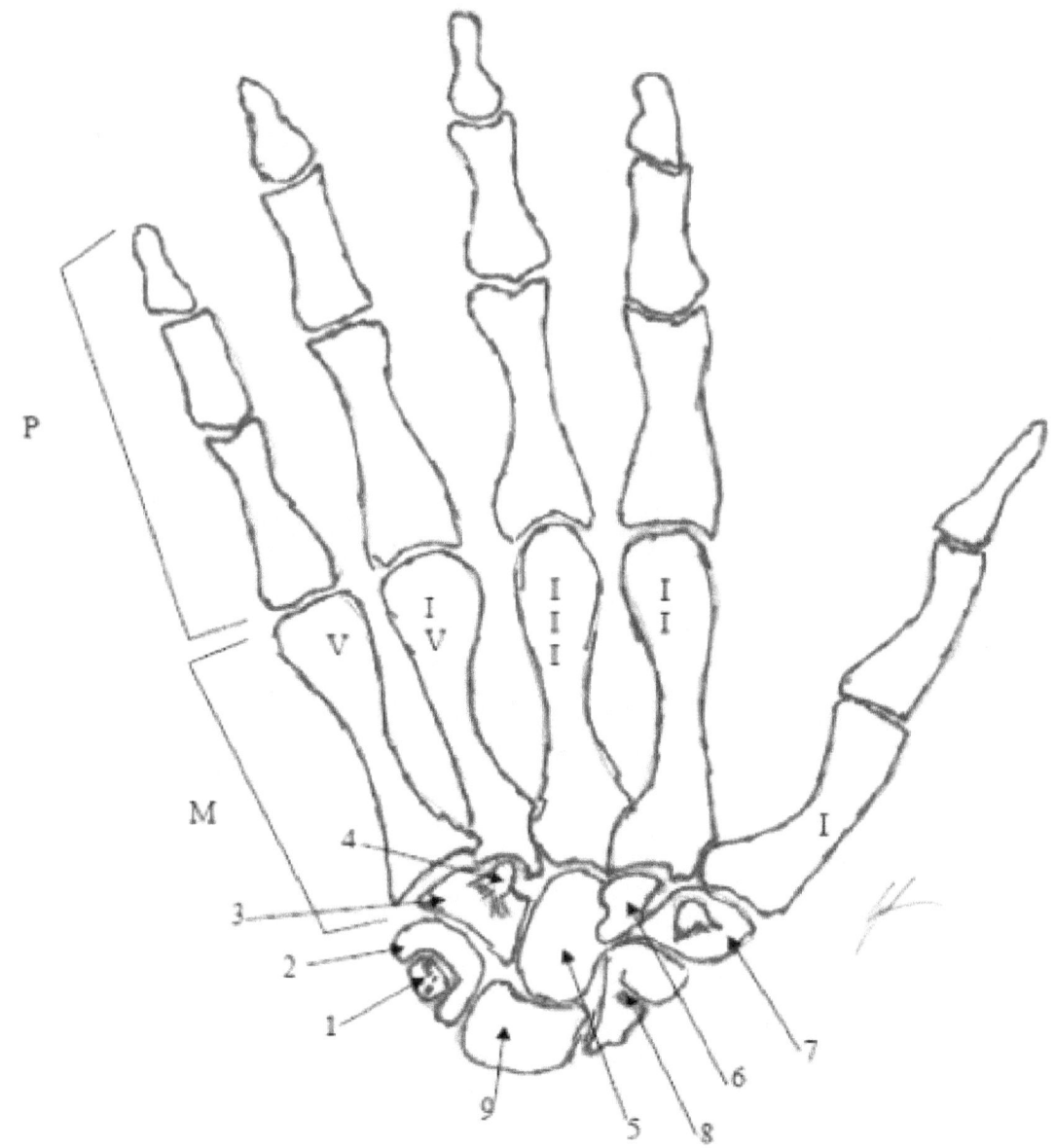

P Phalange
M Metacarpal
   (Roman Numeral I to V for
   indication of digit 1 to digit 5)
1 Pisiform
2 Triquetrum
3 Hamate
4 Hook of Hamate

5 Capitate
6 Trapezoid
7 Trapezium
8 Scaphoid
9 Lunate

Musculoskeletal: Skeleton 10-Lower Extremity Bones

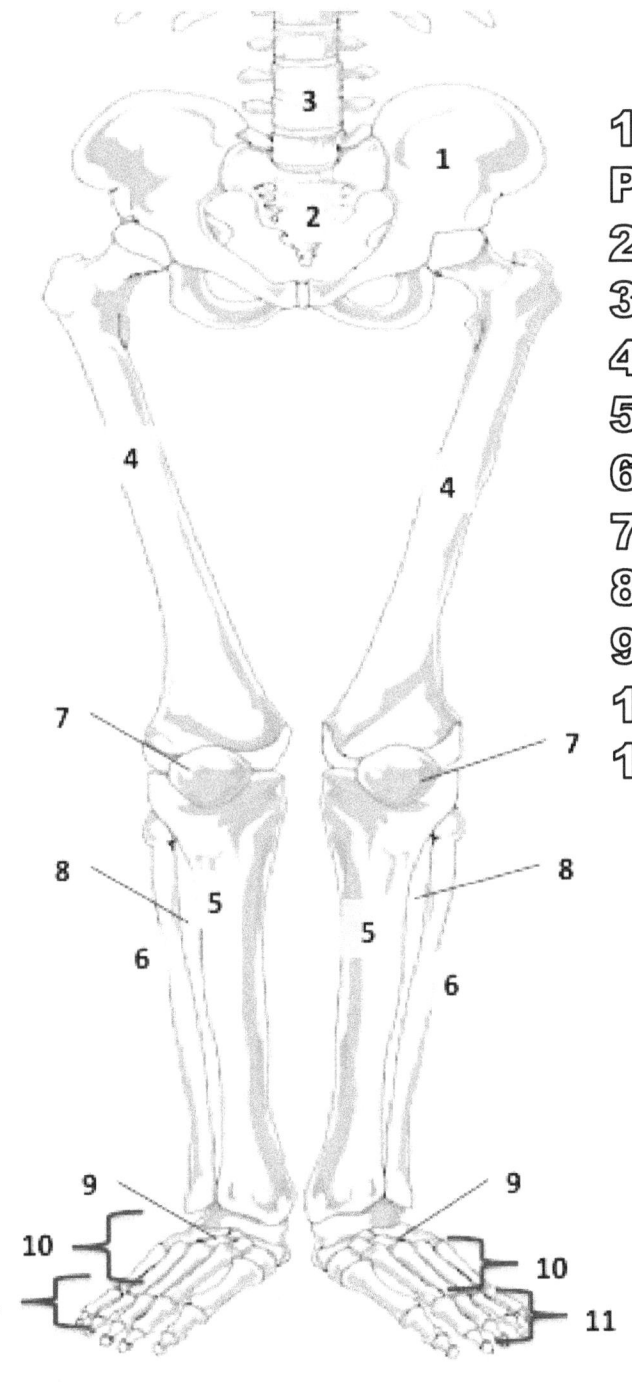

1 Cox (Ilium, Ischium, Pubis)
2 Sacral/Coccygeal
3 Lumbar Vertebrae
4 Femur
5 Tibia
6 Fibula
7 Patella
8 Interosseous Membrane
9 Tarsals
10 Metatarsals
11 Phalanges

Musculoskeletal: Skeleton 11- Pelvic Bones Identification & Landmarks

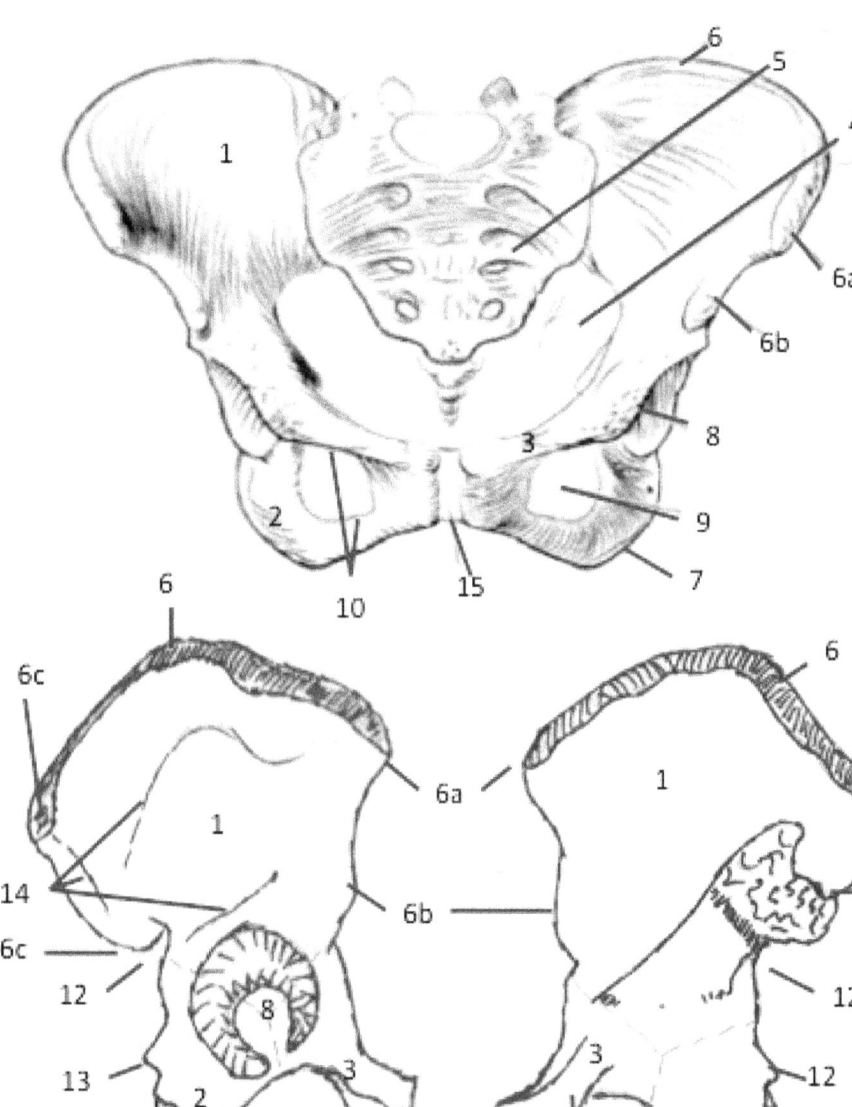

1 Ilium
2 Ischium
3 Pubis
4 Pelvic Bowl Opening
5 Sacrum and Coccygeal
6 Iliac Crest
6a Anterior Superior Iliac Spine
6b Anterior Inferior Iliac Spine
6c Posterior Superior Iliac Spine
6d Anterior Inferior Iliac Spine
7 Ischial Tuberosity
8 Acetabulum
9 Obturator Foramen
10 Ramus of Pubis
11 Pubic Symphasis
12 Sciatic Notch
13 Spine of Ischium
14 Gluteal Lines

Musculoskeletal: Skeleton 12- Lower Extremity Bones Identification & Landmarks 1 (Femur)

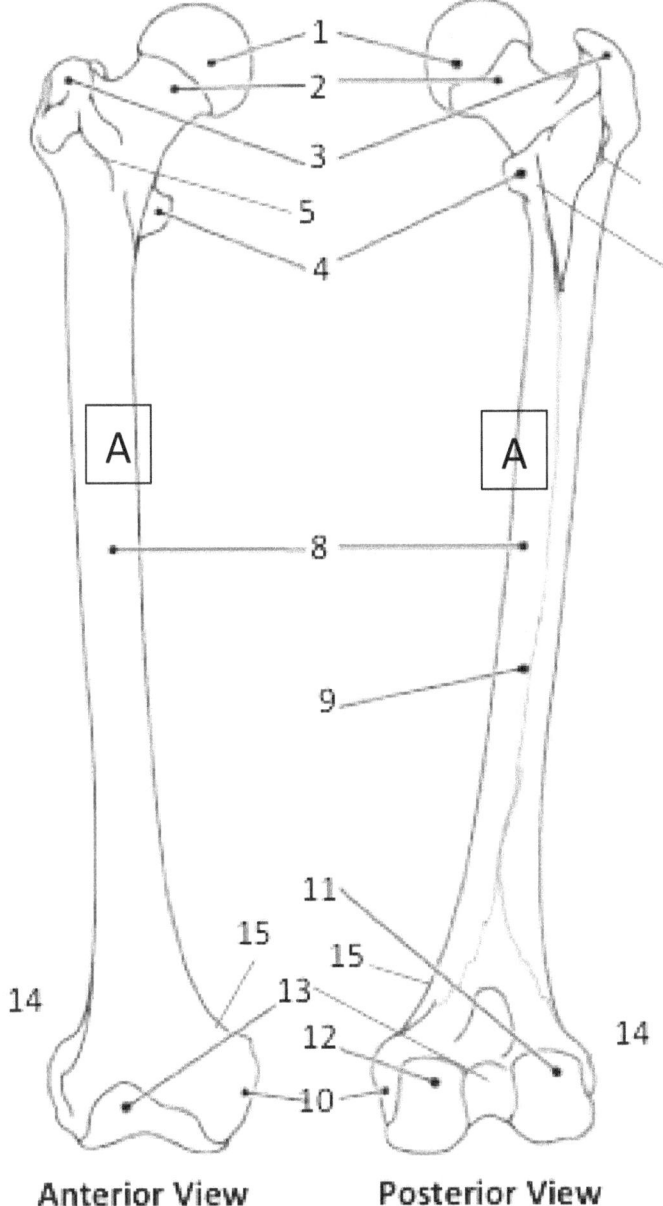

Anterior View   Posterior View

A Femur
1 Head of Femur
2 Neck of Femur
3 Greater Trochanter
4 Lesser Trochanter
5 Intertrochanteric Line (only on anterior)
6 Pectineal Line (only on posterior)
7 Gluteal Line (only on posterior)
8 Body of Femur
9 Linea Aspera (only on posterior)
10 Medial Epicondyle of Femur
11 Lateral Condyle of Femur
12 Medial Condyle of Femur
13 Femoral (Patellar) Groove
14 Lateral Epicondyle of Femur
15 Adductor Tubercle

Musculoskeletal: Skeleton 12- Lower Extremity Bones Identification & Landmarks 2 (Tibia and Fibula)

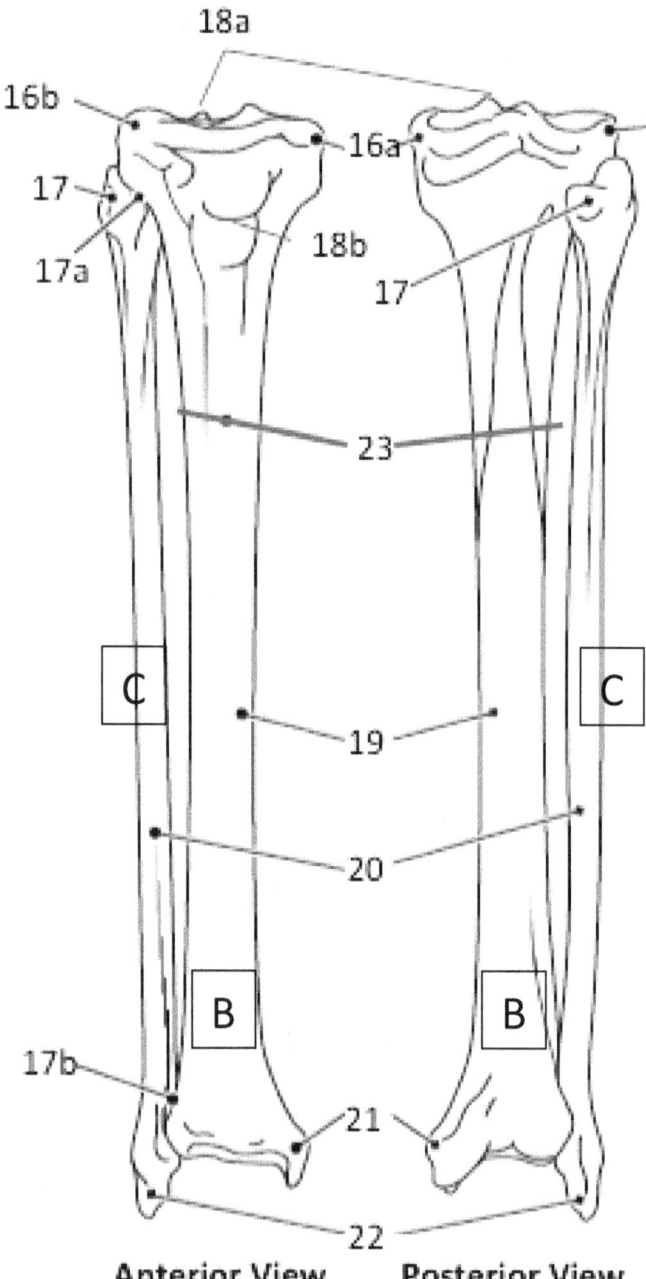

Anterior View   Posterior View

B Tibia
C Fibula

16a Medial Condyle of Tibia
16b Lateral Condyle of Tibia
17 Head of Fibula
17a Proximal Tibofibular junction
17b Distal Tibofibular junction
18a Crown of Tibia (Spurs of ACL and PCL)
18b Tibial Tuberosity
19 Body of Tibia
20 Body of Fibula
21 Medial (Tibial) Malleolus
22 Lateral (Fibular) Malleolus
23 Interosseous Membrane

Musculoskeletal: Skeleton 13- Foot (Tarsals, Metatarsals)

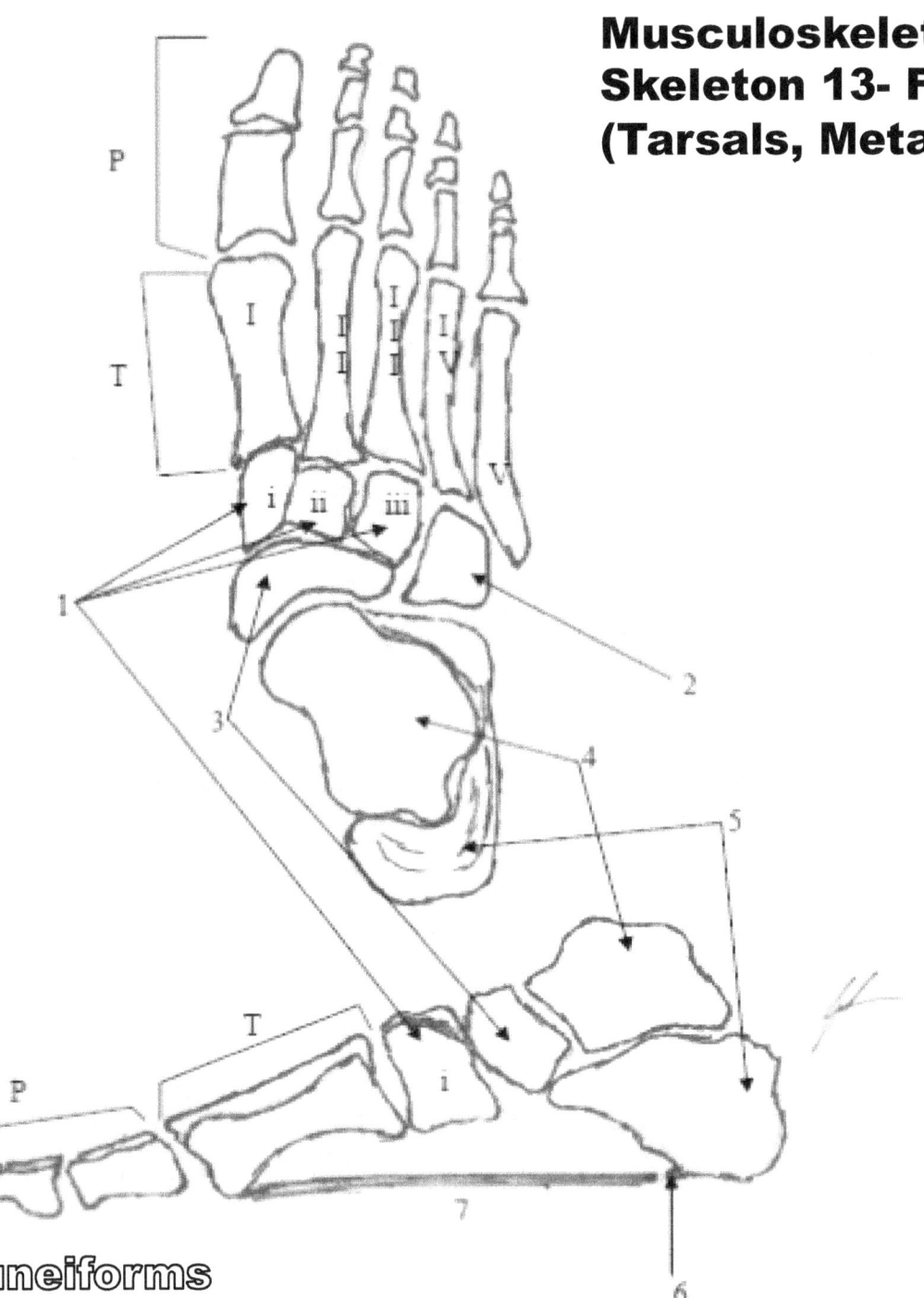

1 Cuneiforms
 (i: Medial, ii: Middle,
 iii: Lateral)
2 Cuboid
3 Navicular
4 Talus
5 Calcaneous
6 Hook of Calcaneous
7 Plantar Fascia

P=Phalanges
T=Tarsal
 (roman numeral I to V
 indicates digits 1 to 5
 for the foot)

Skeletal Muscle Overview

Skeletal muscle is tissue that provides the body with the means to produce locomotion. It does this action through the interaction between ACTIN **and** MYOSIN within the MUSCLE CELL pulling on the tendon at the INSERTION of the muscle into a bone bringing that bone closer to the non-moving ORIGIN of the muscle. When it comes to skeletal muscles, form follows function. Meaning that the adaptations in the in the structure of the muscle will be based on the functionality of the muscle and desired level of force that the muscle must produce. In which the basic shape of the muscle and associated connective tissue will dictate the strength of muscular action. The strength of the muscle depends on the muscle's general shapes are determined by the connective tissue that is interwoven within the skeletal muscle and provides anchorage to the bone and thus provides the means for movement to occur. As it relates to the muscle action, the most influential classification is the arrangement of the connective tissue within the muscle, these patterns include: PARALLEL, CONVERGENT, PENNATE, and CIRCULAR.

Also, we can discuss muscles as being active in the direction we want to move AGONIST or against the movement we want ANTAGONIST. There are three distinct types of contractions that a muscle goes through to make the movements we want as being CONCENTRIC, or a shortening of the muscle, ECCENTRIC, the lengthening of the muscle, and ISOMETRIC, contraction with no seen movement of the muscle.

When discussing the nomenclature of muscle anatomy, there are several general means to name a muscle. These methods for naming a muscle will describe either the location of the muscle, the number of origins (or muscle units) within the muscle, the arrangement of the connective tissue fibers within the muscle, the relative size of the muscle within a group of muscles, the shape of the muscle, and/or the action of the muscle.

Musculoskeletal: Muscle-Structures 1

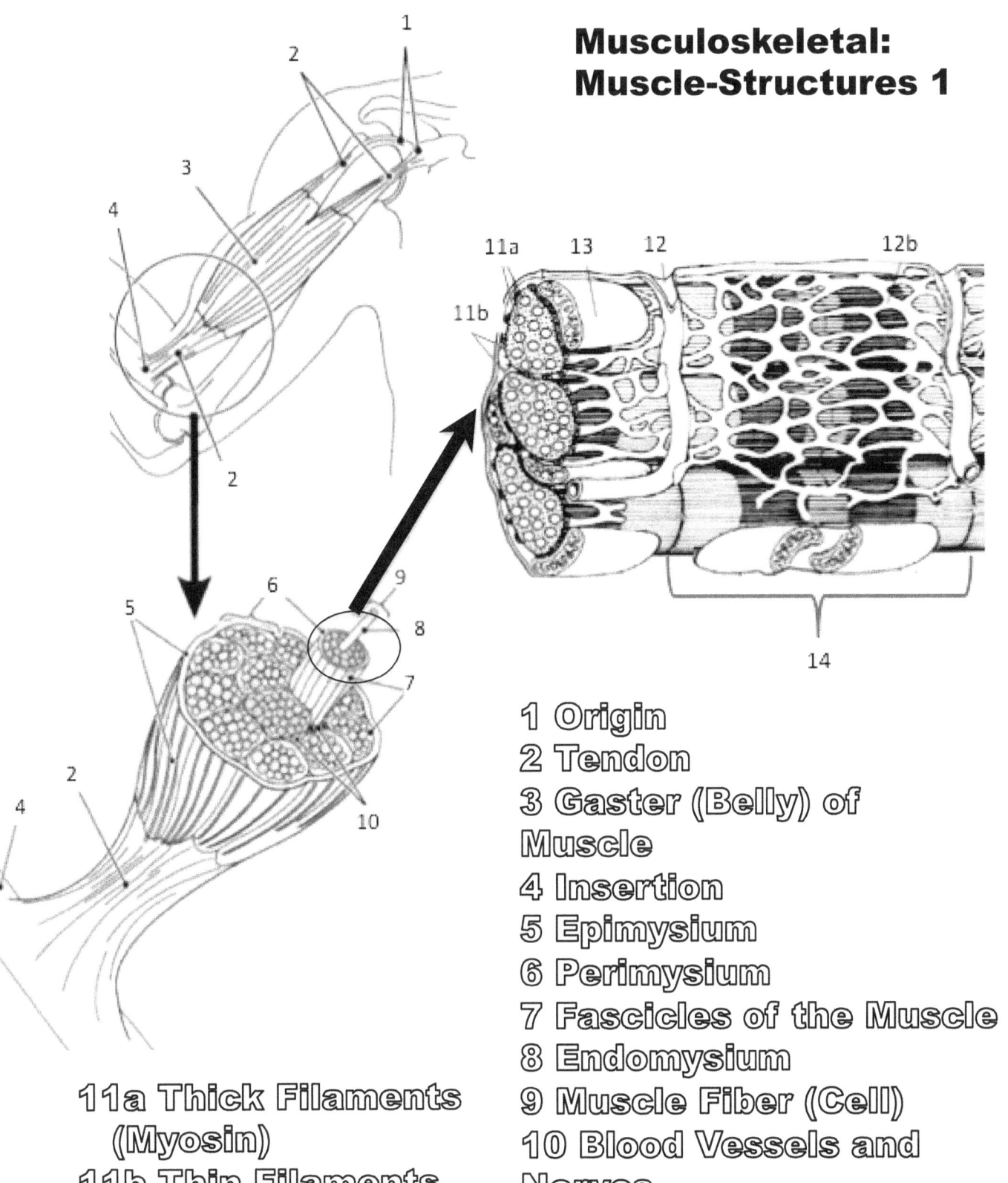

1 Origin
2 Tendon
3 Gaster (Belly) of Muscle
4 Insertion
5 Epimysium
6 Perimysium
7 Fascicles of the Muscle
8 Endomysium
9 Muscle Fiber (Cell)
10 Blood Vessels and Nerves
11a Thick Filaments (Myosin)
11b Thin Filaments (Actin)
12 Triad of Transverse Tubules
12b Sarcoplasmic Reticulum
13 Mitochondria
14 Sarcomere

Musculoskeletal: Muscle Structure 2- Tissue Structures (Sarcomere)

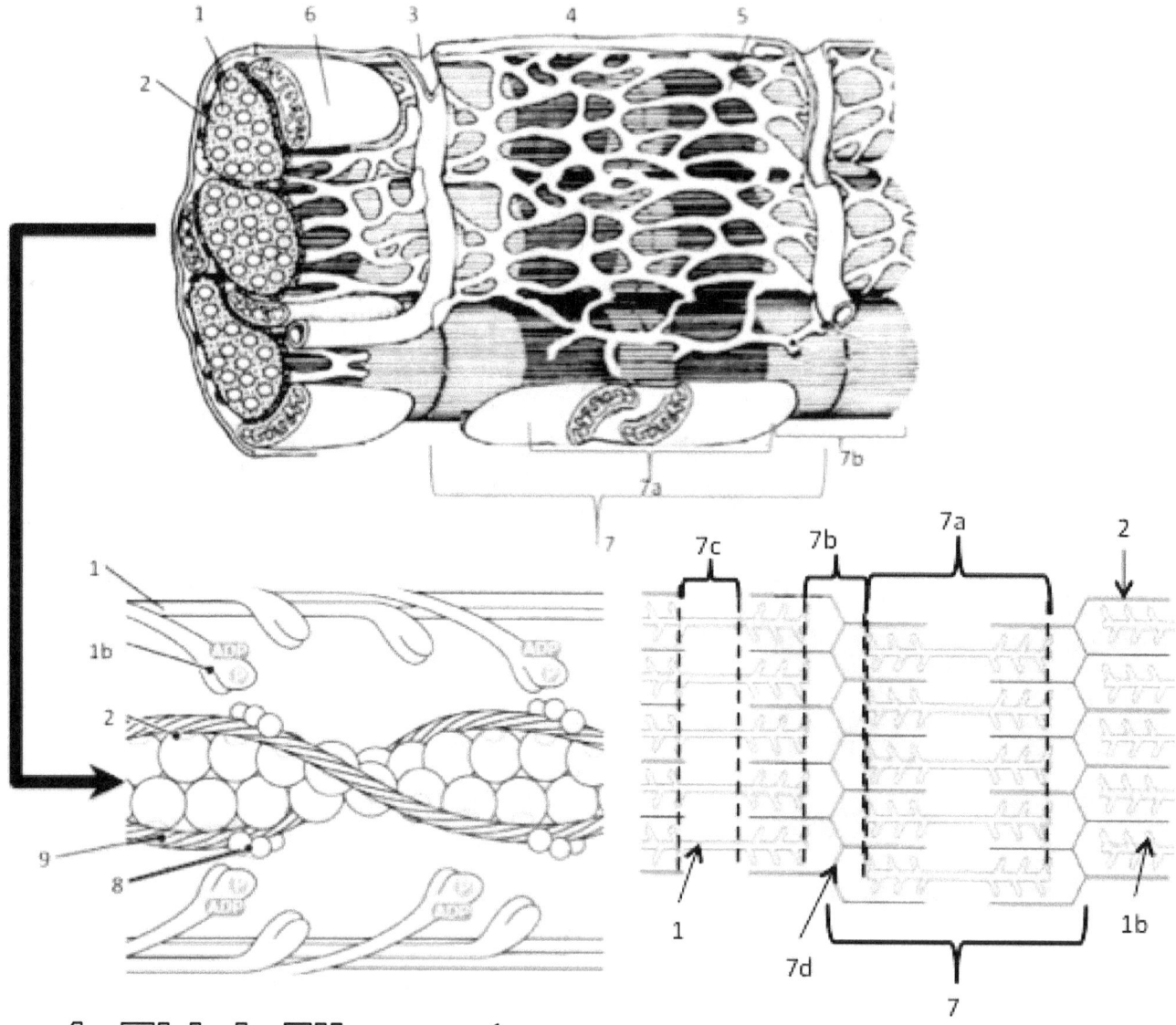

1=Thick Filaments (Myosin)
1a=Head of Myosin
2=Thin Filaments (Actin)
3= Triad of Transverse Tubules
4=Transverse Tubules
5=Sarcoplasmic Reticulum
6=Mitochondria
7=Sarcomere
7a=A-band
7b= I-Band
7c=M-Zone
7d=Z-line
8=Troponin
9=Tropomyosin

Musculoskeletal: Muscle Structure 3- Neuromuscular Junction

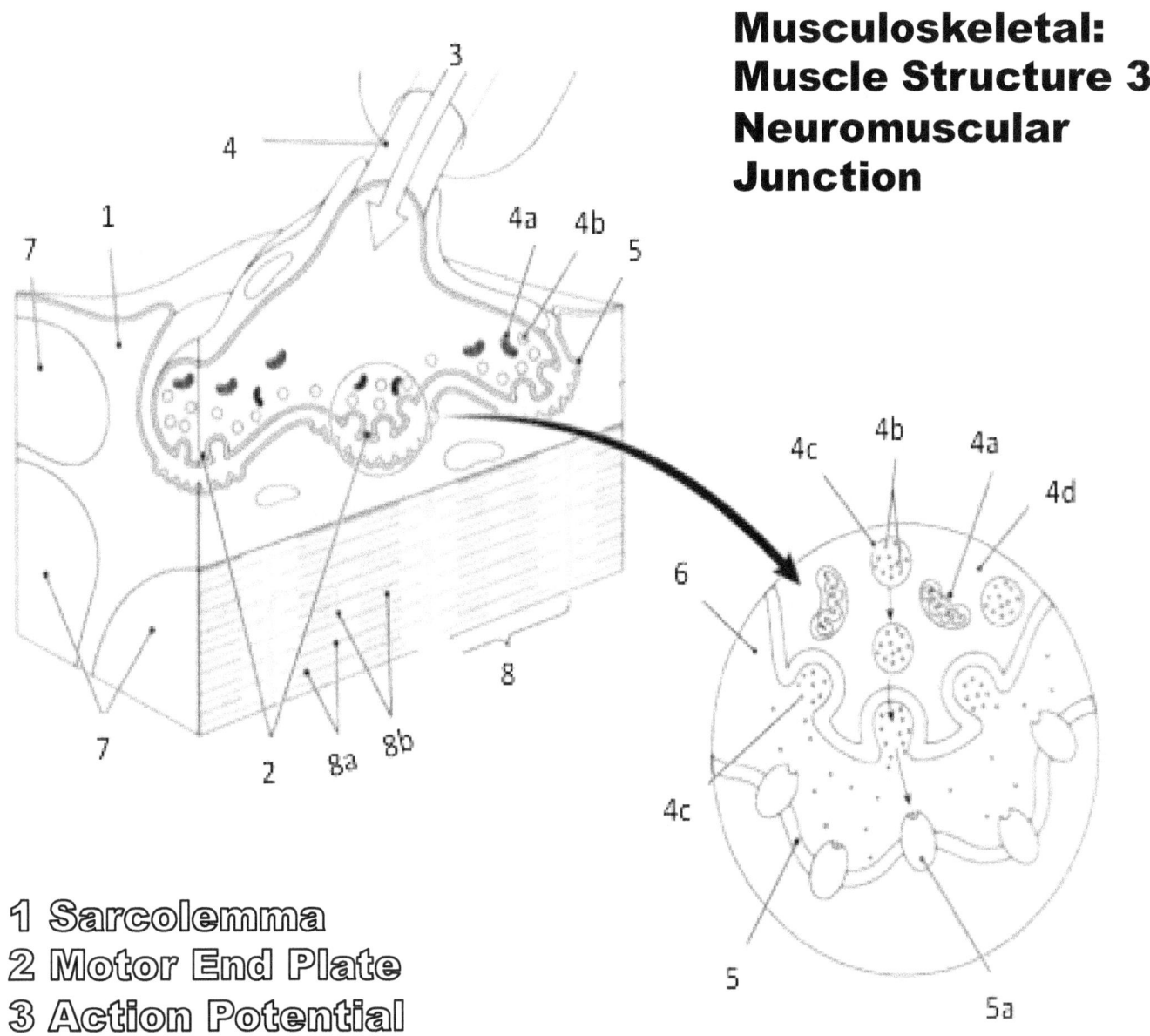

1 Sarcolemma
2 Motor End Plate
3 Action Potential
4 Motor Neuron
4a Mitochondria
4b Vesicle of Neurotransmitters
4c Neurotransmitter (Acetylcholine)
4d Synaptic Button (Terminal)
5 Sarcolemma of Motor End Plate
5a Acetylcholine Receptors
6 Synaptic Cleft of Motor End Plate
7 Muscle Fibrils
8 Sarcomere
8a Thin Filament (Actin Protein)
8b Thick Filament (Myosin Protein)

Musculoskeletal: Muscle 1- Overview (Anterior)

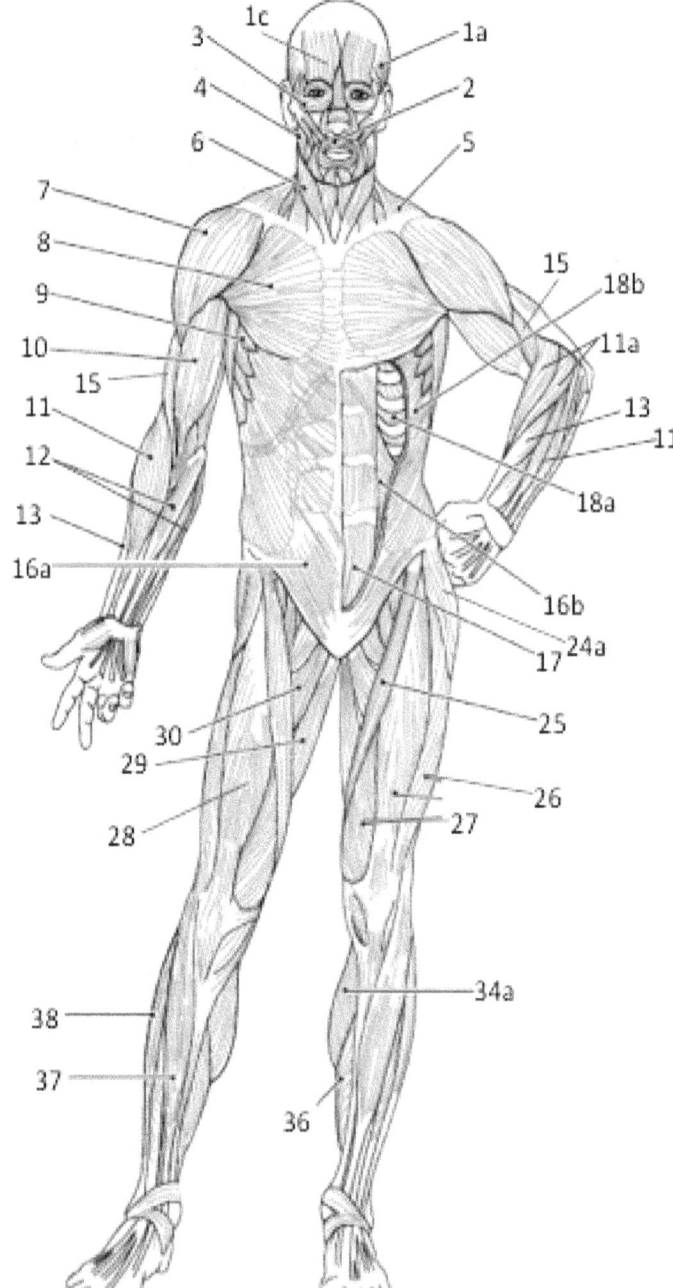

10=Biceps Brachii
11=Brachioradialis
11a=Extensor Carpi Radialis Longus and Brevis
11b=Common Extensors
12=Common Flexors
13=Extensor Carpi Radialis
14=Triceps Brachii
14a=Triceps Tendon
15=Brachialis
16a=External Abdominal Oblique
16b=Internal Abdominal Oblique
17=Rectus Abdominis
18a=Internal Intercostal
18b=External Intercostal
24a=Tensor Fascia Latte
25=Sartorius
26=Vastus Lateralis
27=Vastus Medialis
28=Rectus Femoris
29=Adductor Magnus
30=Adductor Longus
34a=Medial Head Gastrocnemius
36=Soleus
 (Deep to Tendon on posterior)
37=Tibialis Anterior
38=Peroneal Muscles

1a=Temporalis
1c=Frontalis
2=Orbicularis Oris
3=Orbicularis Oculi
4=Masseter
5=Trapezius
6=Sternocleidomastoid
7=Deltoid
8=Pectoralis Major
9=Serratus Anterior

Musculoskeletal: Muscle 2- Overview (Posterior)

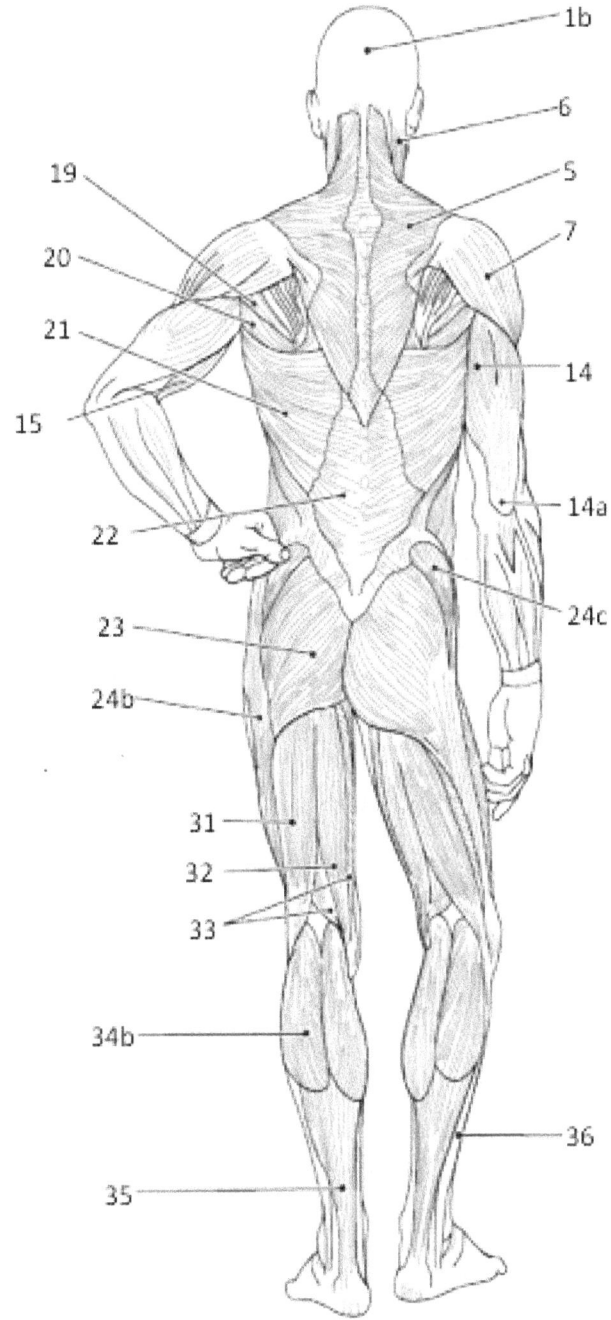

1b Epicranial Aponeurosa
5 Trapezius
6 Sternocleidomastoid
7 Deltoid
14 Triceps Brachii
14a Triceps Tendon
15 Brachialis
19 Teres Minor
20 Teres Major
21 Latissimus Dorsi
22 Thoracolumbar Fascia
23 Gluteus Maximus
24b Iliotibial Band
24c Gluteus Medius
25 Sartorius
31 Biceps Femoris
32 Semitendinosus
33 Semimembranosus
34a Medial Head Gastrocnemius
34b Lateral Head Gastrocnemius
35 Calcaneal (Achilles) Tendon
36 Soleus
   (Deep to Tendon on posterior)
38 Peroneal Muscles

Musculoskeletal: Muscle 3- Head & Neck

1 Frontalis
2 Temporalis
3 Nasalis
4 Orbicularis Oris
5 Zygomaticus
6 Mentalis
6a Quadratus Labii Inferioris
7 Orbicularis Oculi
8 Quadratus Labii Superioris
9 Masseter
10 Triagnularis
11 Sternocleidomastoid
12 Trapezius
13 Buccinator
14 Digastricus
15 Epicranial Aponeurosa

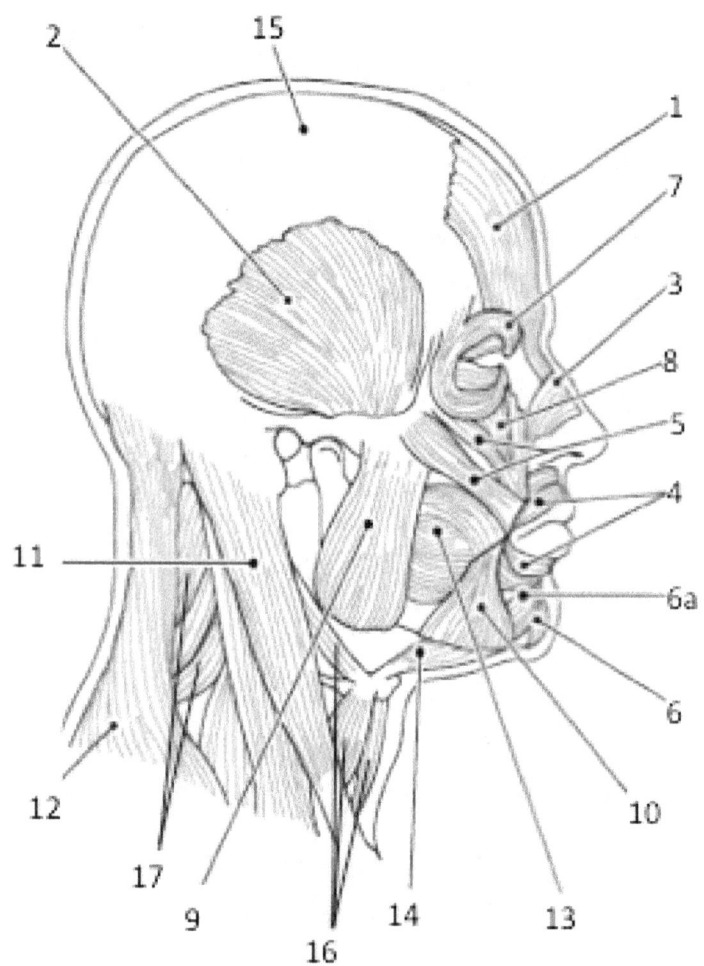
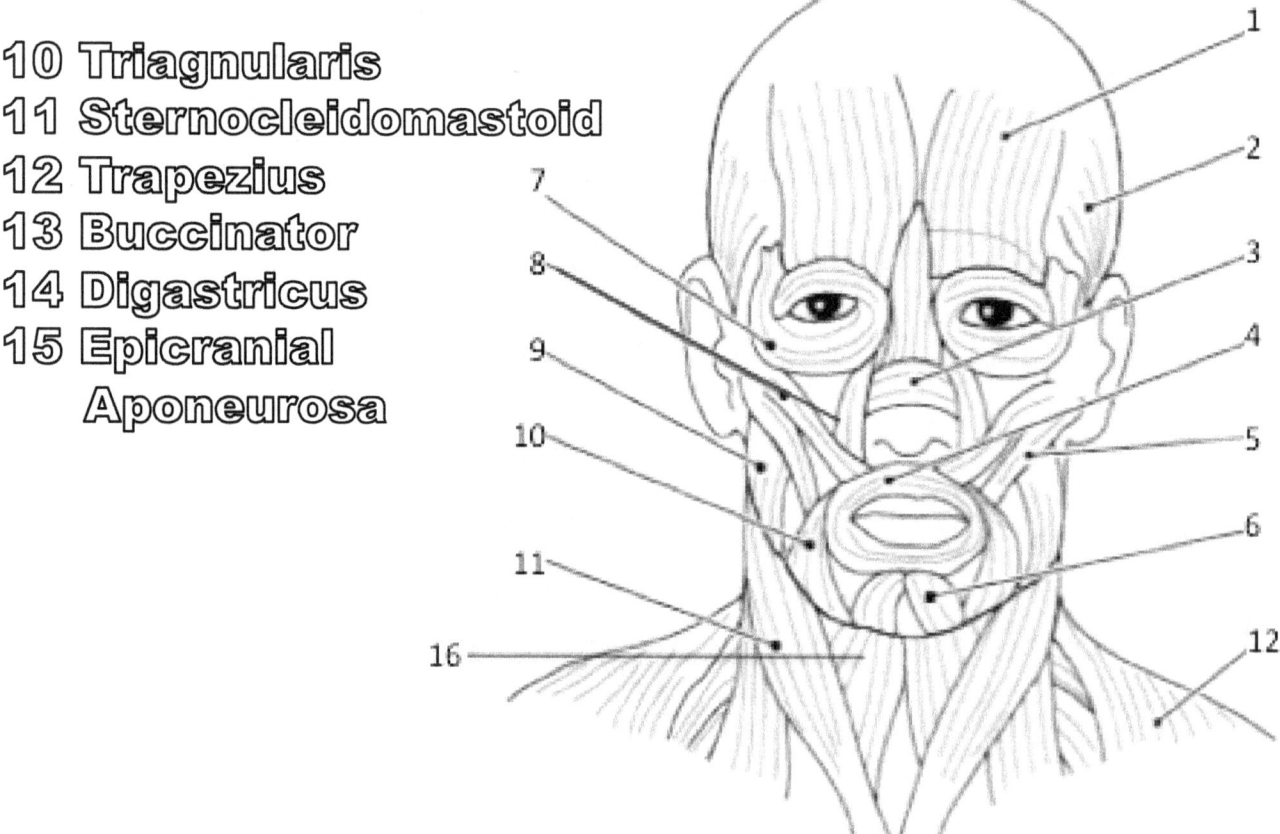

Musculoskeletal: Muscle 4- Upper Extremity

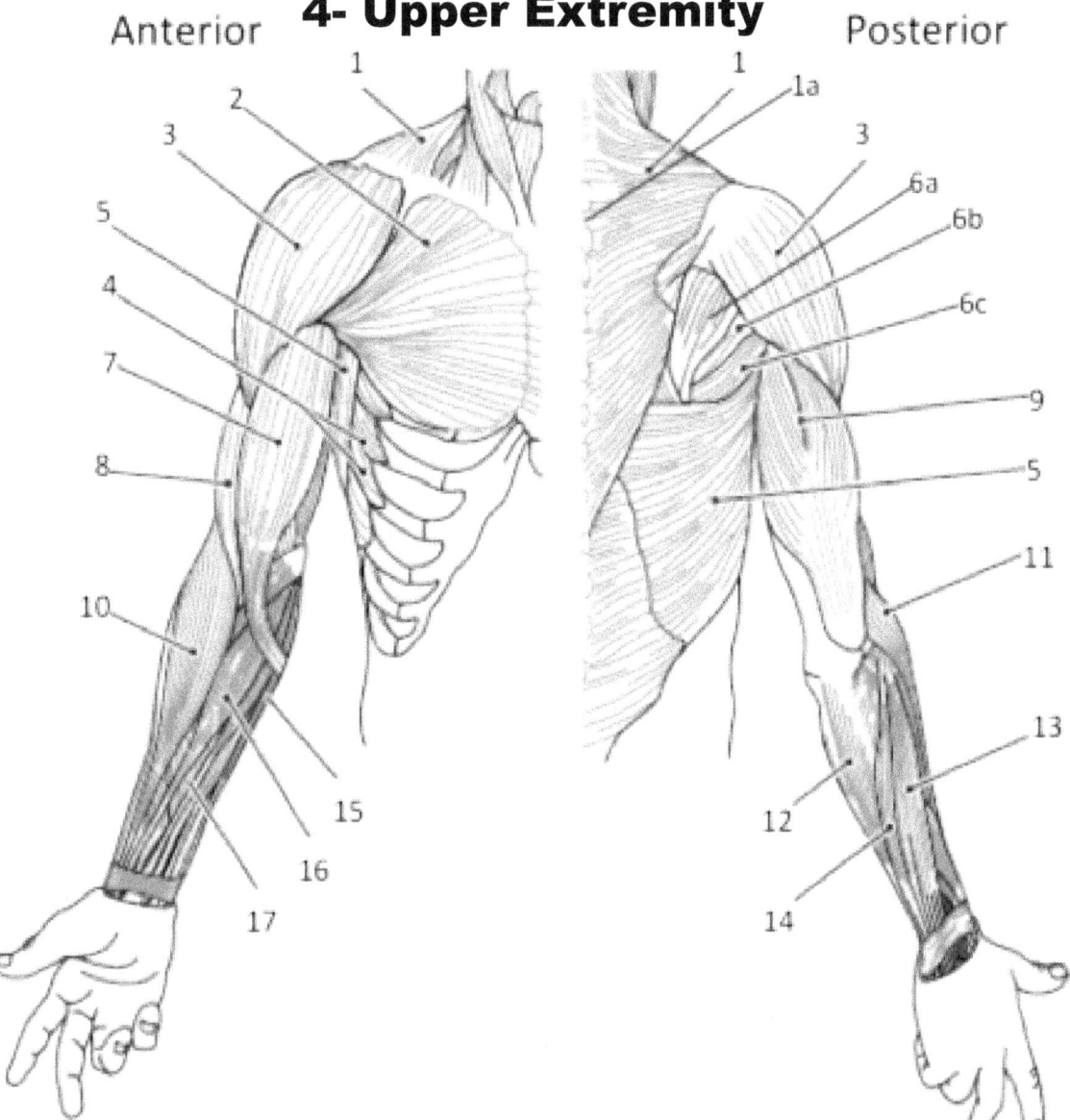

1=Trapezius
1a=Ligamentum Nuchae
2=Pectoralis Major
3=Deltoid
4=Serratus Anterior
5=Latissimus Dorsi
6a=Infraspinatus
6b=Teres Minor
6c=Teres Major
7=Biceps Brachii
8=Brachialis
9=Triceps Brachii
10=Brachioradialis
11=Extensor Carpi Radialis Longus
12=Extensor Carpi Ulnaris
13=Extensor Digitorum
14=Extensor Digiti Minimi
15=Flexor Carpi Ulnaris
16=Flexor Digitorum Superficialis
17=Flexor Carpi Radialis

Musculoskeletal: Muscle 5- Abdominal/Trunk Muscle

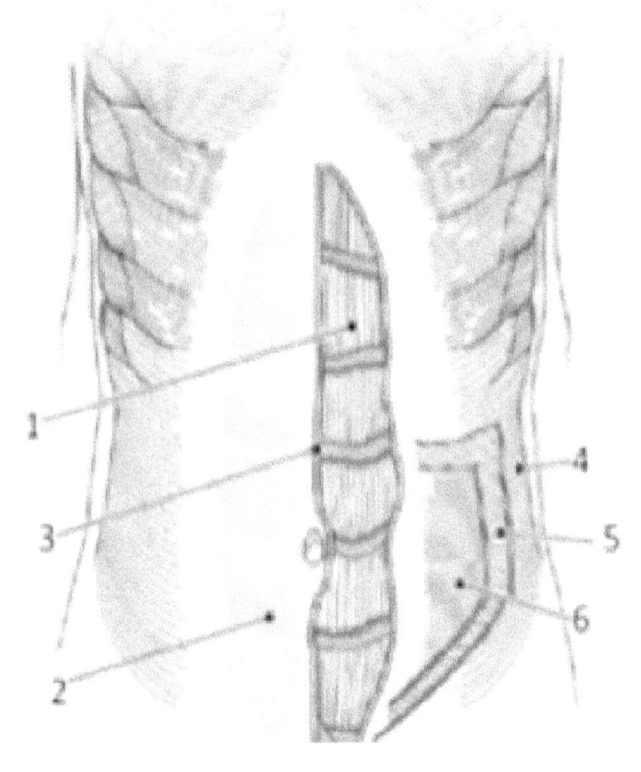

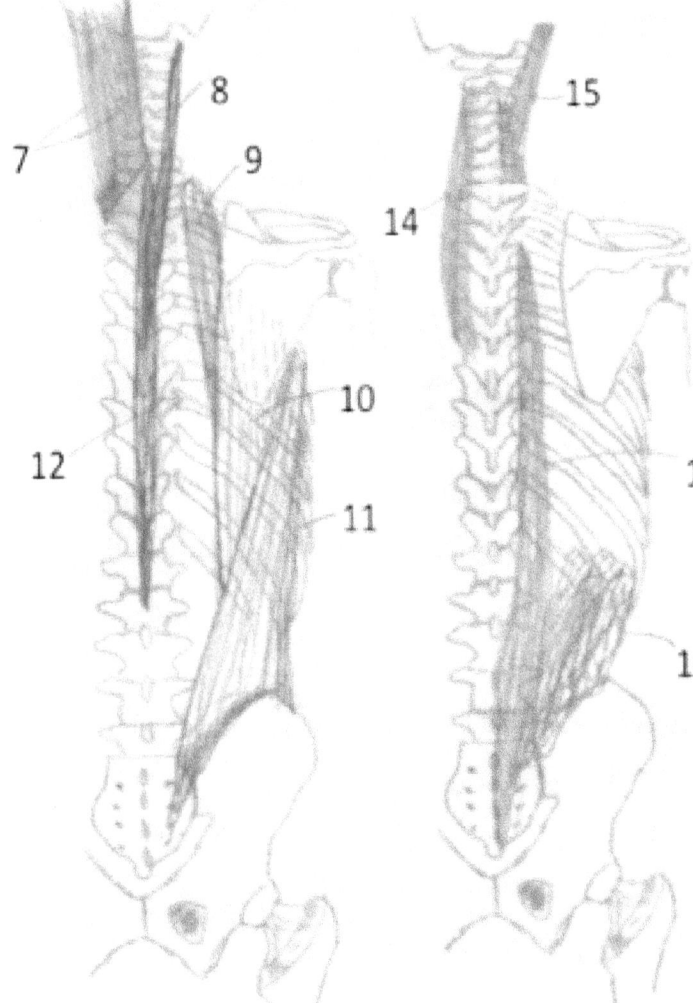

1 Rectus Abdominis
2 Abdominal Sheath
3 Linea Alba
4 External Abdominal Oblique
5 Internal Abdominal Oblique
6 Transvers Abdominus
7 Splenius Capitis
8 Splenius Cervicis
9 Iliocostalis Cervicis
10 Iliocostalis Thoracis
11 Iliocostalis Lumborum
12 Splenius Thoracis
13 Longissimus Thoracis
14 Longissimus Cervicis
15 Longissimus Capitis
16 Quadratus Lumborum

Musculoskeletal: Muscle 6- Lower Extremity (Anterior)

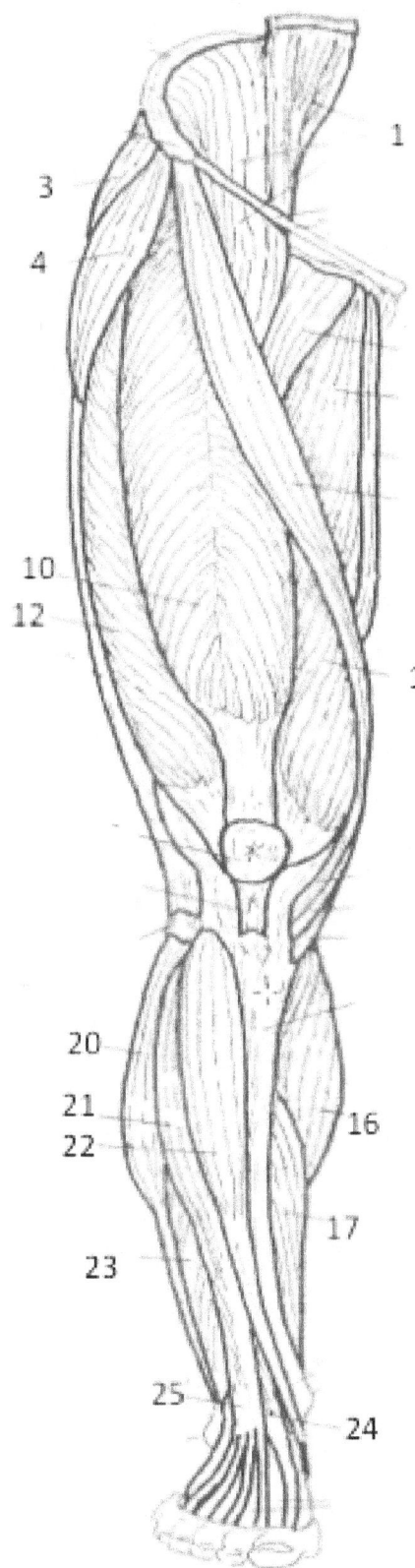

Anterior View

1 Iliopsoas
3 Gluteus Medius
4 Tensor Fascia Lata
5 Adductor Brevis
6 Adductor Magnus
7 Gracilis
8 Sartorius
10 Rectus Femoris
11 Vastus Medialis
12 Vastus Lateralis
16 Gastrocnemius
17 Soleus
20 Peroneus Longus
21 Extensor Digitorum Longus
22 Tibialis Anterior
23 Peroneus Tertius
24 Extensor Hallucis
25 Peroneus Brevis
26 Tibialis Posterior

Musculoskeletal: Muscle 6- Lower Extremity (Posterior)

2 Gluteus Maximus
3 Gluteus Medius
3a Iliotibial Band
4 Tensor Fascia Lata
5 Adductor Brevis
6 Adductor Magnus
7 Gracilis
8 Sartorius
9 Semitendinosus
13 Semimembranosus
14 Biceps Femoris
15 Plantaris
16 Gastrocnemius
17 Soleus
18 Calcaneal (Achilles) Tendon
19 Flexor Digitorum Longus
20 Peroneus Longus
26 Tibialis Posterior

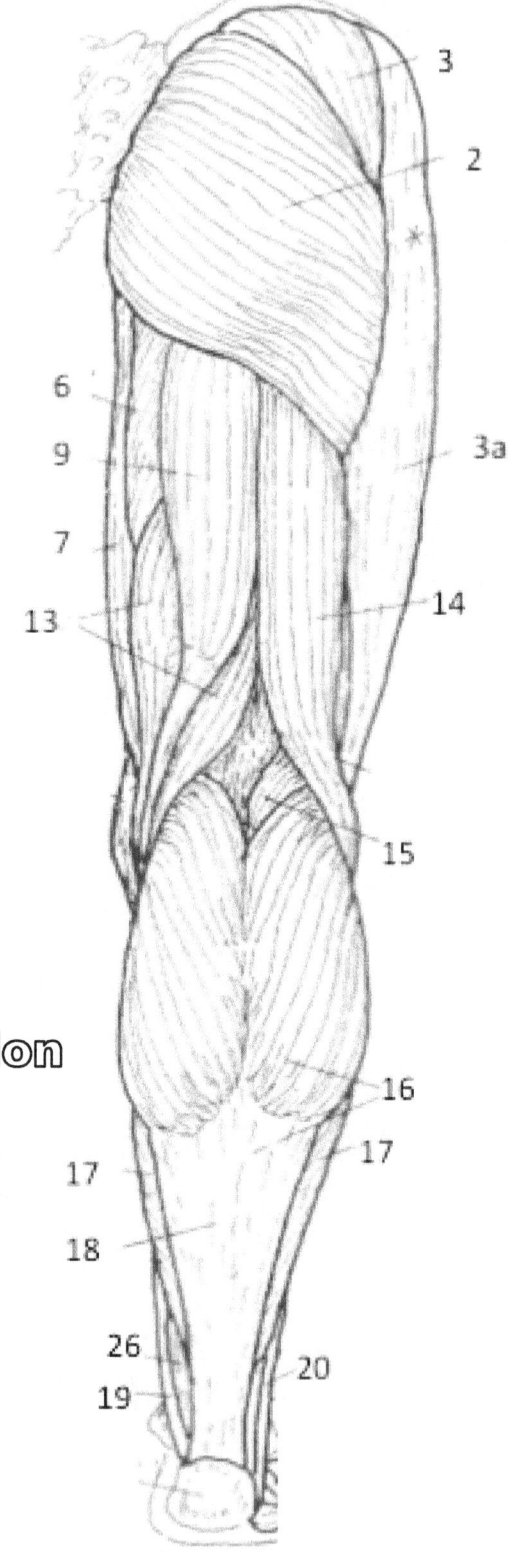

Posterior View

Nervous System

Neurons, Nerves and Structures of the Cerebral Cortex

Overview

The nervous system is a network of specialized cells (NEURONS) that are designed to receive and transmit information from one area of the body to another area and support cells (GLIAL). It is important to remember that while we talk a lot about "NERVES" when discussing cellular actions neurons are not the same as nerves. Nerves are actually a collection of various types of neurons and support cells and not specifically a neuron.

The method by which neurons pass the information is though a voltage change of their membrane (ACTION POTENTIAL) that involves a sequence of opening and closing "voltage-gated channels" along the membrane by either CONTINUOUS (all of the membrane) or SALTATORY (only selected regions) conduction. And then pass the signal from one cell to another cell by specialized chemical messengers (NEUROTRANSMITTERS) that will open specialized transmembrane channels (receptors) that will then induce another action potential or inhibit the production of an action potential.

Neurons and nerves organize in such a way that there are systems of neurons, CENTRAL (CEREBRAL CORTEX & SPINAL CORD) and PERIPHERAL (NERVES), that form response (REFLEX ARCS) through converging, diverging, or reverberating networks within the nervous system. This organization begins with the reception of stimulus and ending with the COGNITIVE integration of signals to form a MEMORY or develop a COGNITIVE RESPONSE (A PURPOSED MOVEMENT) in response to the stimulus in such a way that quick maintenance of homeostasis within the body. Building overlapping networks in such a way that are there are actually multiple nervous systems within the nervous system as a whole and each of these systems are integrated to maintain the homeostasis within the human body.

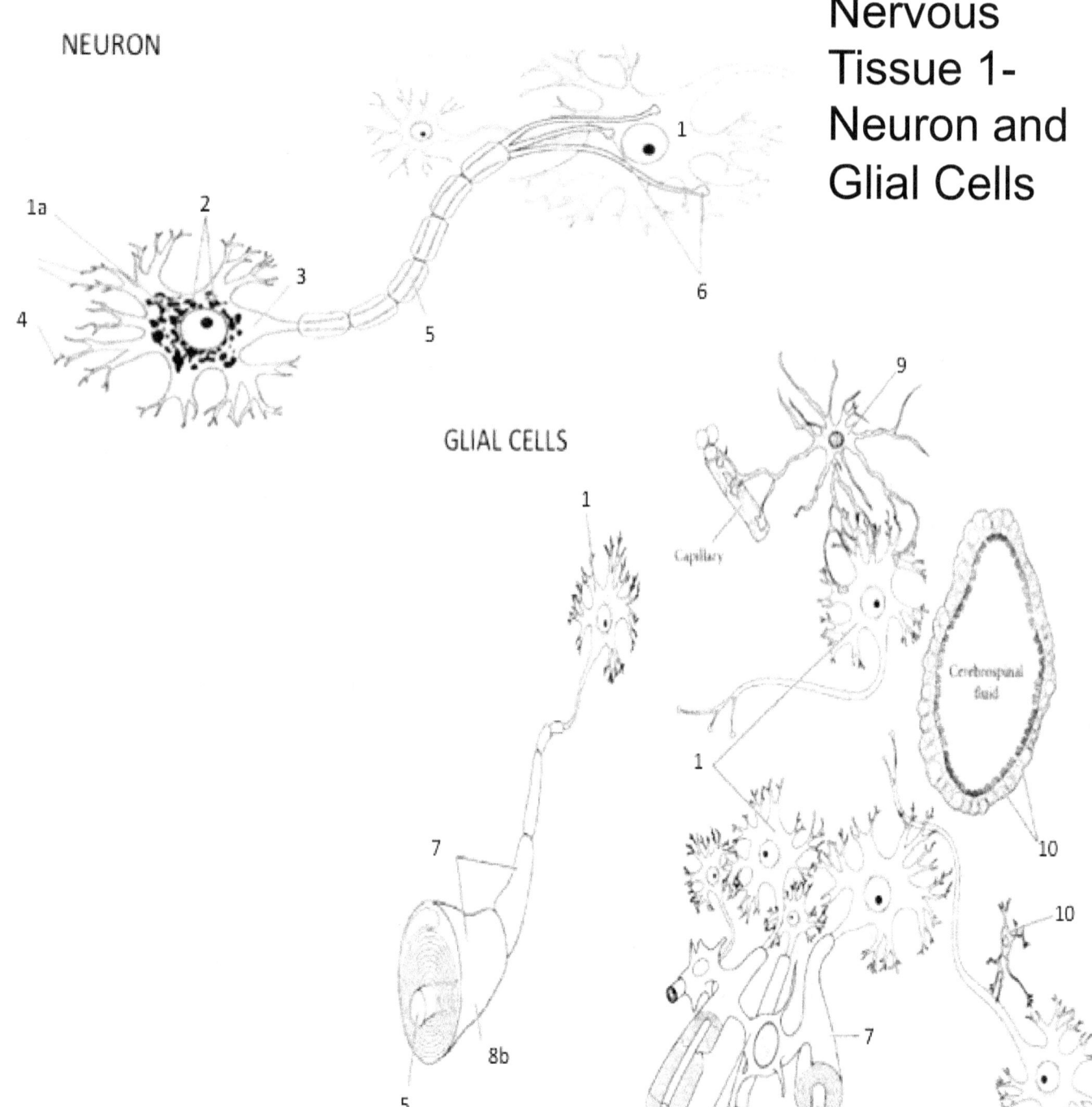

Nervous Tissue 1- Neuron and Glial Cells

1 Cell Body of Neuron
1a Nucleus
2 Nussle Bodies
3 Axon Hillock
4 Dendrites
5 Axon
6 Axon Terminals/ Terminal Branches
7 Myelin
8a Oligodendrocytes
8b Schwann Cells
9 Astrocytes
10 Microglial cells

Nervous Tissue 2- Synapse & Action Potential

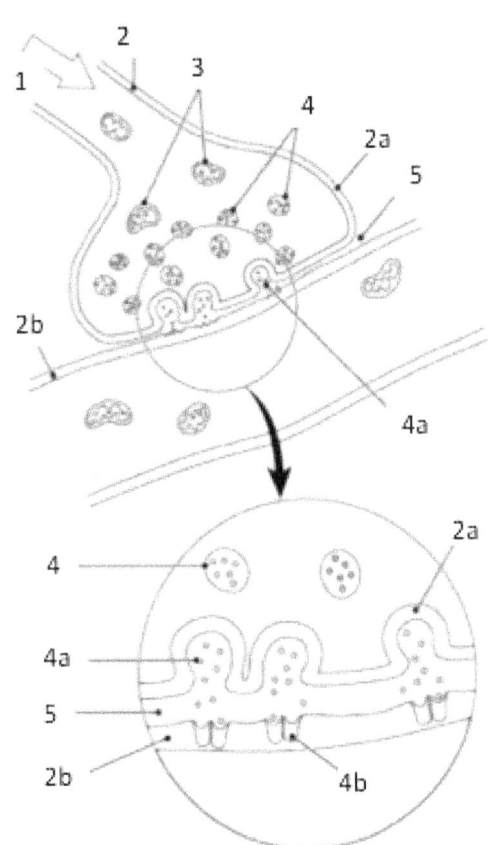
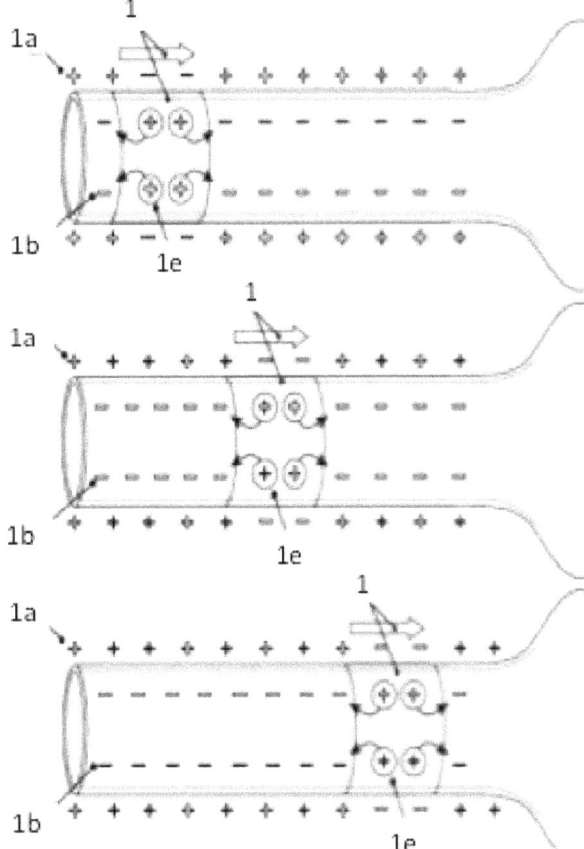

1 Action Potential
1a Positive "Charges"
1b Negative "Charges"
1d Myelin
1e Region of charge "spread"
1f Saltatory conduction
2 Terminal Branch
2a Pre-synaptic Membrane
2b Post-Synaptic Membrane
3 Mitochondria
4 Vesicles of Neurotransmitters
4a Neurotransmitter
4b Receptor for Neurotransmitter
5 Synaptic Cleft

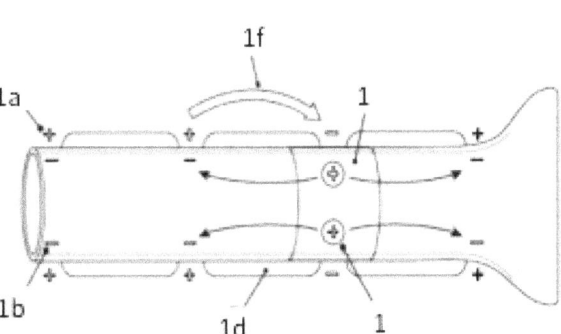

Nervous Tissue 3- Action Potential

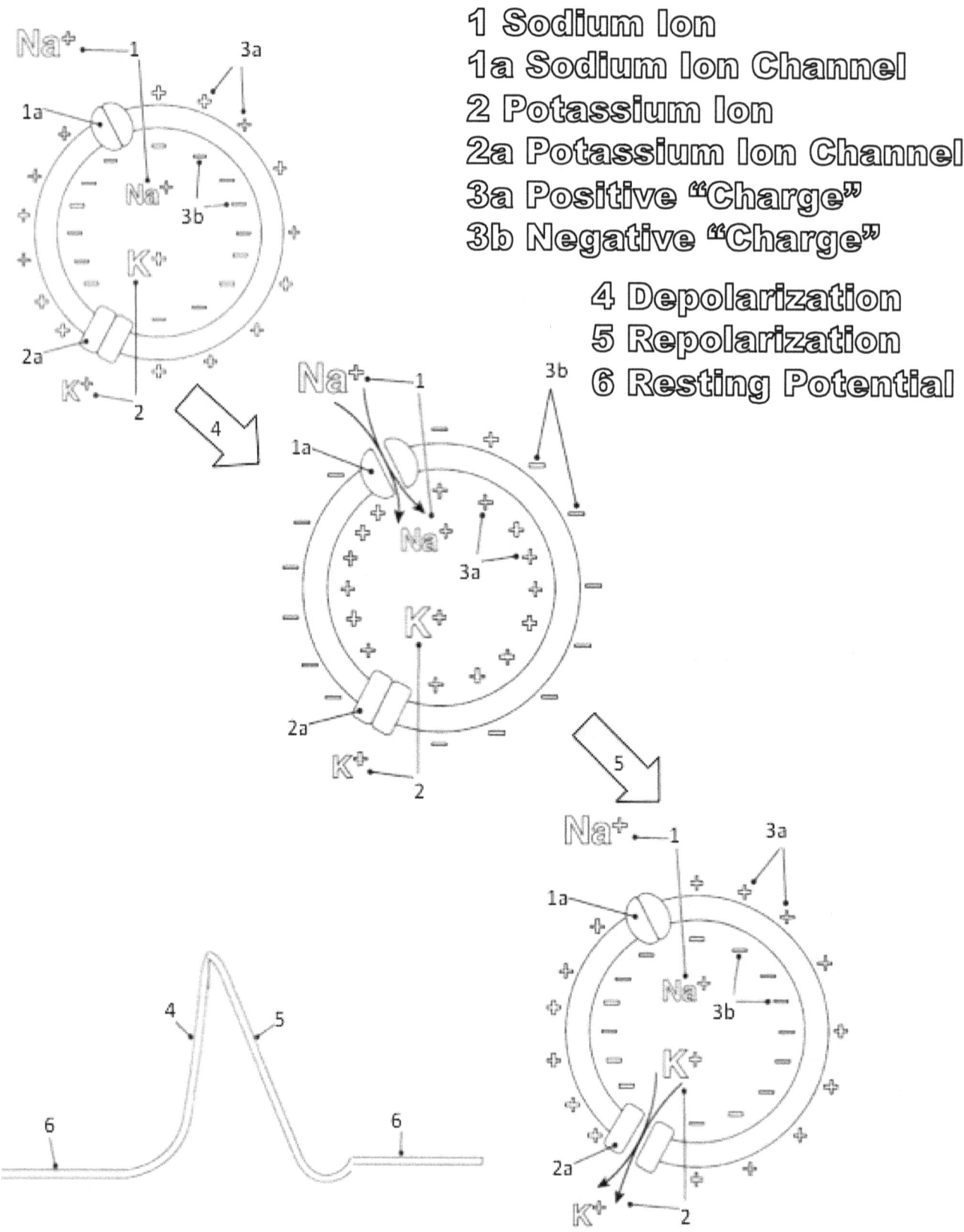

1 Sodium Ion
1a Sodium Ion Channel
2 Potassium Ion
2a Potassium Ion Channel
3a Positive "Charge"
3b Negative "Charge"

4 Depolarization
5 Repolarization
6 Resting Potential

Cerebral Cortex 1 - Lateral & Medial View

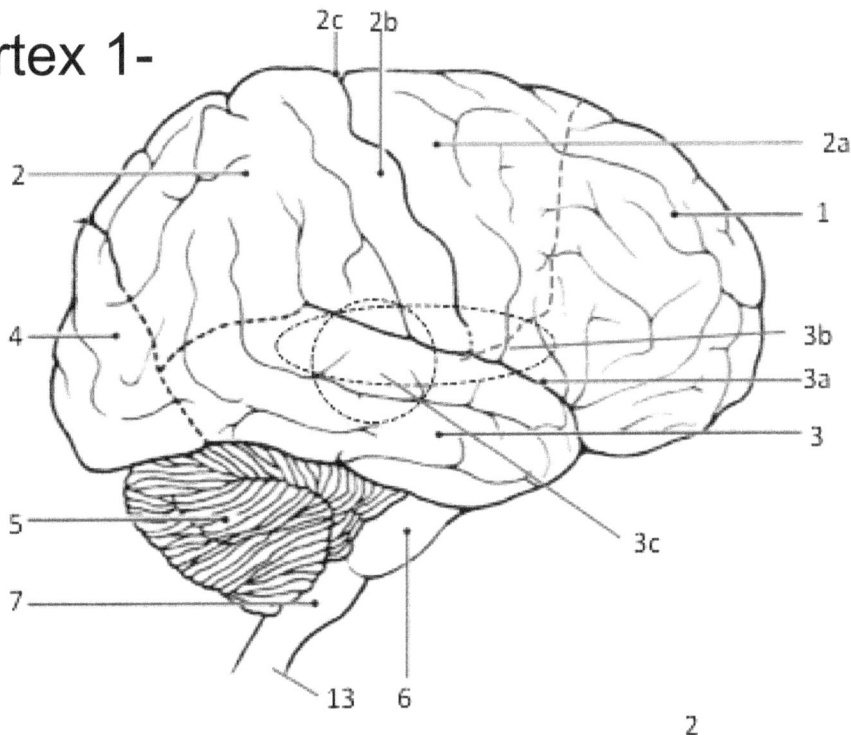

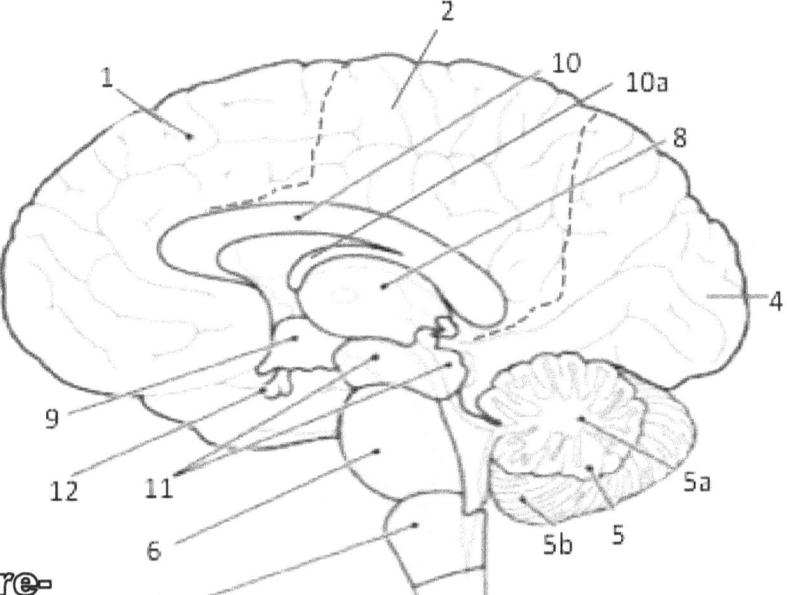

1 Frontal Lobe
2 Parietal Lobe
2a Primary Motor Cortex (Pre-Central Gyrus)
2b Primary Somatosensory Cortex (Post-Central Gyrus)
2c Central Sulcus
3 Temporal Lobe
3a Lateral Sulcus
3b Wernicke's Area
3c Brocca's Area
4 Occipital Lobe
5 Cerebellum
5a Arbor Vitae of Cerebellum
5b Vermis of Cerebellum
6 Pons
7 Medulla Oblongata
8 Thalamus
9 Hypothalamus
10 Corpus Callosum
10a Fornix
11 Tectum & Limbic System
12 Pituitary Glans
13 Spinal Cord

Cerebral Cortex 2 - Coronal & Inferior View

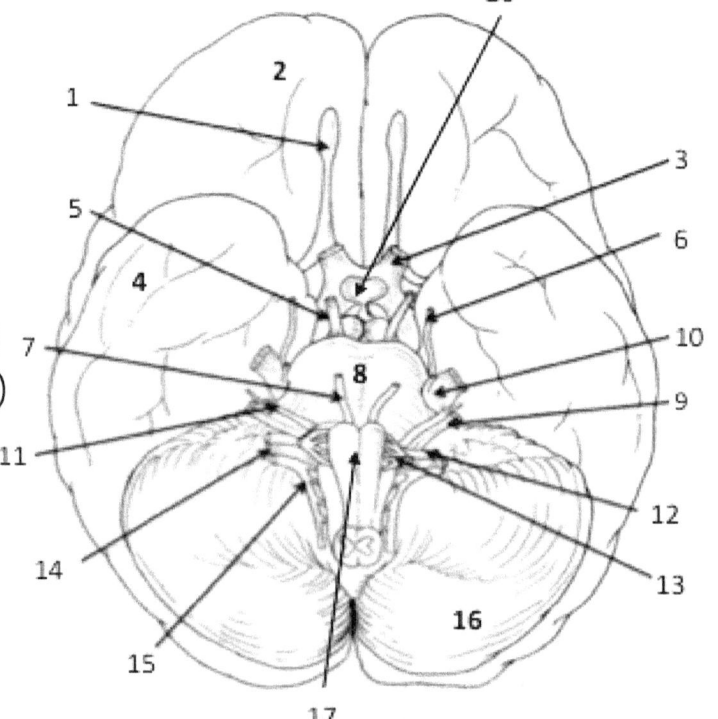

1 Cranial Nerve I (Olfactory)
2 Frontal Lobe
3 Cranial Nerve II (Optic)
4 Temporal Lobe
5 Cranial Nerve III (Oculomotor)
6 Cranial Nerve IV (Trochlear)
7 Cranial Nerve VI (Abducens)
8 Pons
9 Cranial Nerve VIII (Cochlearvestibular)
10 Cranial Nerve V (Trigeminal)
11 Cranial Nerve VII (Facial)
12 Cranial Nerve IX (Glossopharyngeal)
13 Cranial Nerve X (Vagus)
14 Cranial Nerve XI (Spinal Accessory)
15 Cranial nerve XII (Hypoglossal)
15 Cerebellum
17 Medulla Oblongata
18a Putnam
18b Globus Pallidus
19 Caudate Nucleus
20 Hypothalamus
21 Thalamus
22a Cerebral Cortex Axons ("White" Matter)
22b Cerebral Cortex Cellular Layer
22c Corpus Callosum (Commissure Fibers)

Spinal Cord 2-Dermatome & Myotomes

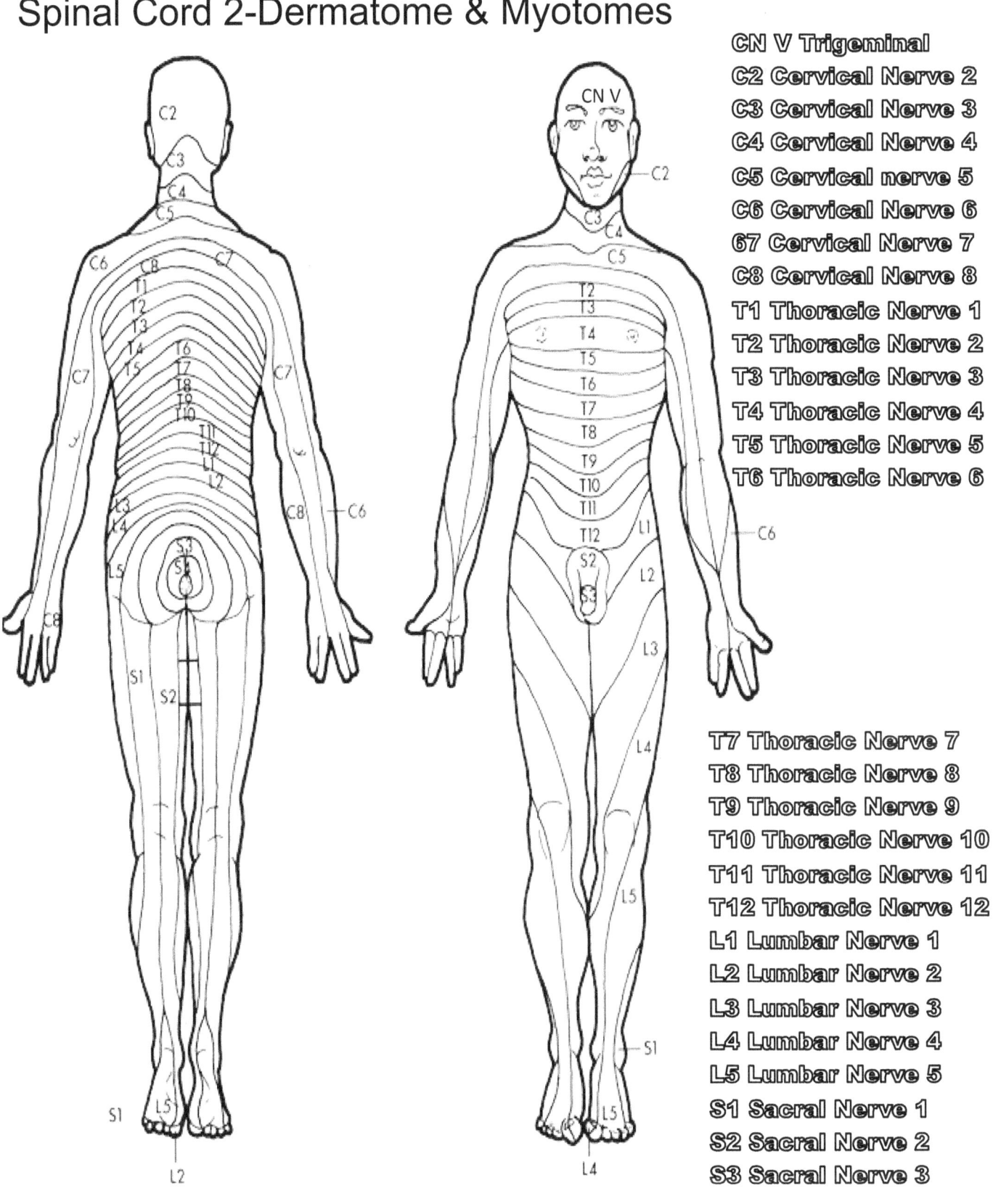

Peripheral Nerves 1 - Brachial Plexus

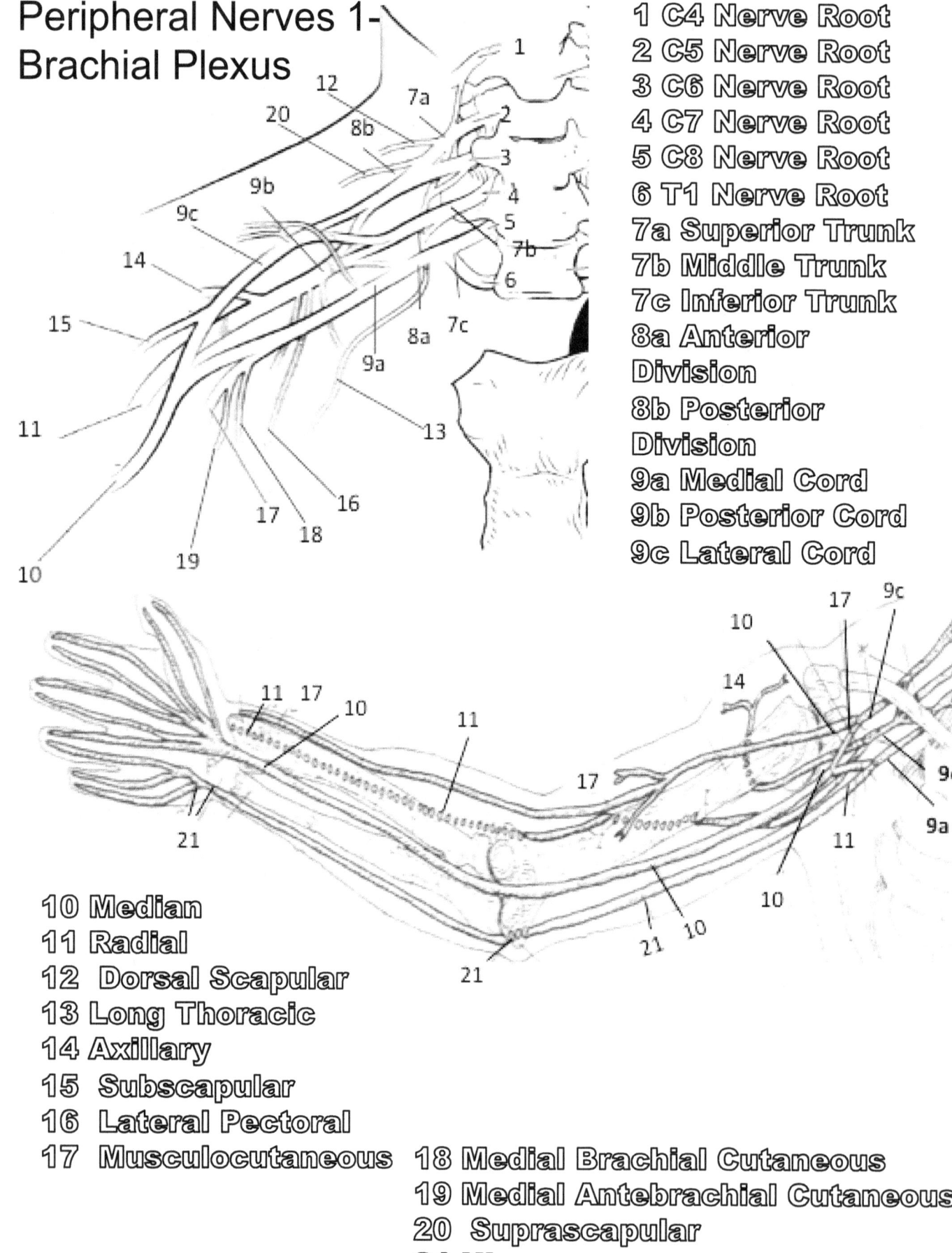

1 C4 Nerve Root
2 C5 Nerve Root
3 C6 Nerve Root
4 C7 Nerve Root
5 C8 Nerve Root
6 T1 Nerve Root
7a Superior Trunk
7b Middle Trunk
7c Inferior Trunk
8a Anterior Division
8b Posterior Division
9a Medial Cord
9b Posterior Cord
9c Lateral Cord
10 Median
11 Radial
12 Dorsal Scapular
13 Long Thoracic
14 Axillary
15 Subscapular
16 Lateral Pectoral
17 Musculocutaneous
18 Medial Brachial Cutaneous
19 Medial Antebrachial Cutaneous
20 Suprascapular
21 Ulnar

Peripheral Nerves 2 - Lumbar/Sacral Plexus

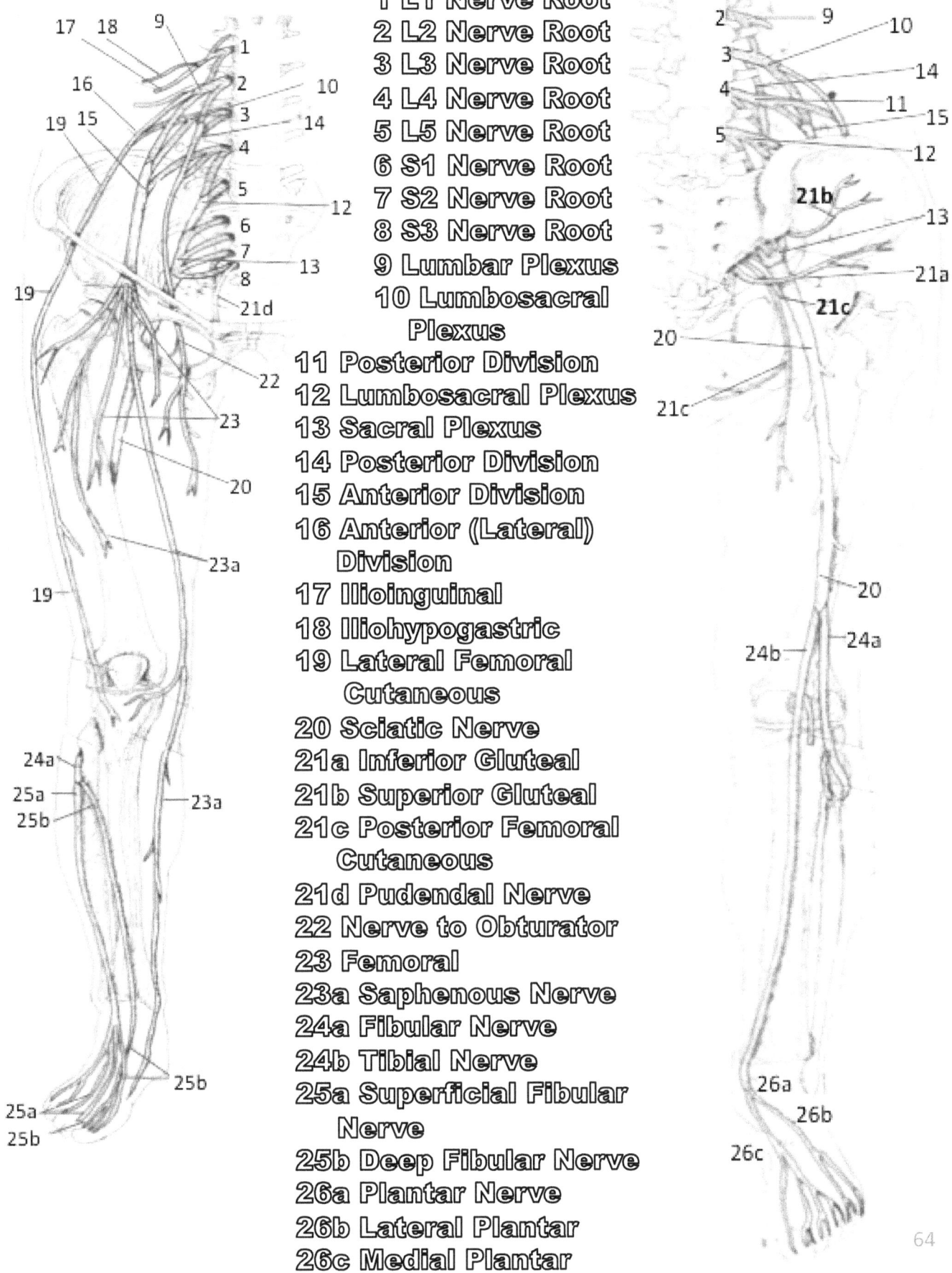

1. L1 Nerve Root
2. L2 Nerve Root
3. L3 Nerve Root
4. L4 Nerve Root
5. L5 Nerve Root
6. S1 Nerve Root
7. S2 Nerve Root
8. S3 Nerve Root
9. Lumbar Plexus
10. Lumbosacral Plexus
11. Posterior Division
12. Lumbosacral Plexus
13. Sacral Plexus
14. Posterior Division
15. Anterior Division
16. Anterior (Lateral) Division
17. Ilioinguinal
18. Iliohypogastric
19. Lateral Femoral Cutaneous
20. Sciatic Nerve
21a. Inferior Gluteal
21b. Superior Gluteal
21c. Posterior Femoral Cutaneous
21d. Pudendal Nerve
22. Nerve to Obturator
23. Femoral
23a. Saphenous Nerve
24a. Fibular Nerve
24b. Tibial Nerve
25a. Superficial Fibular Nerve
25b. Deep Fibular Nerve
26a. Plantar Nerve
26b. Lateral Plantar
26c. Medial Plantar

Special Senses 1- Olfaction & Gustation

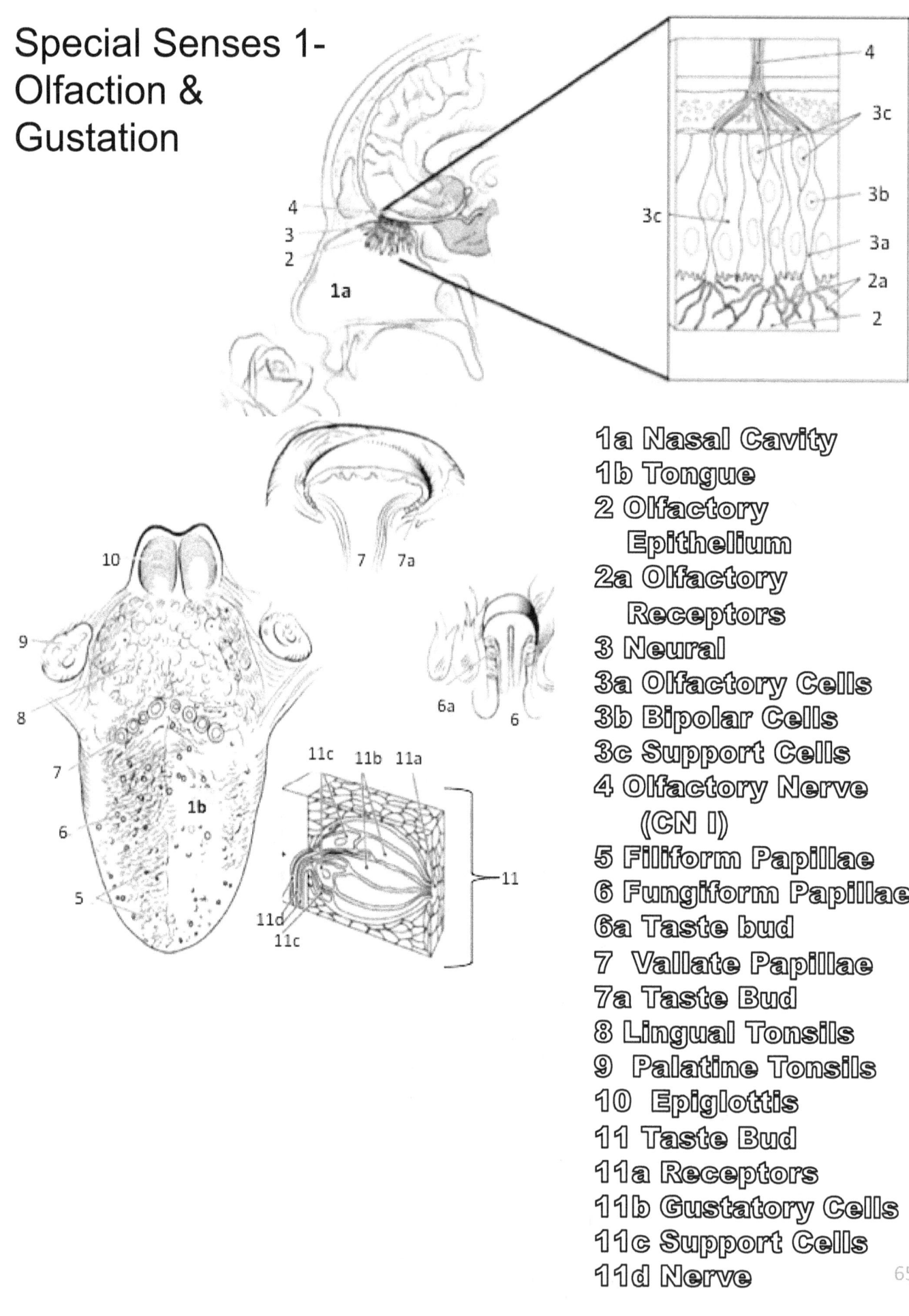

1a Nasal Cavity
1b Tongue
2 Olfactory Epithelium
2a Olfactory Receptors
3 Neural
3a Olfactory Cells
3b Bipolar Cells
3c Support Cells
4 Olfactory Nerve (CN I)
5 Filiform Papillae
6 Fungiform Papillae
6a Taste bud
7 Vallate Papillae
7a Taste Bud
8 Lingual Tonsils
9 Palatine Tonsils
10 Epiglottis
11 Taste Bud
11a Receptors
11b Gustatory Cells
11c Support Cells
11d Nerve

Special Senses 2 - Auditory

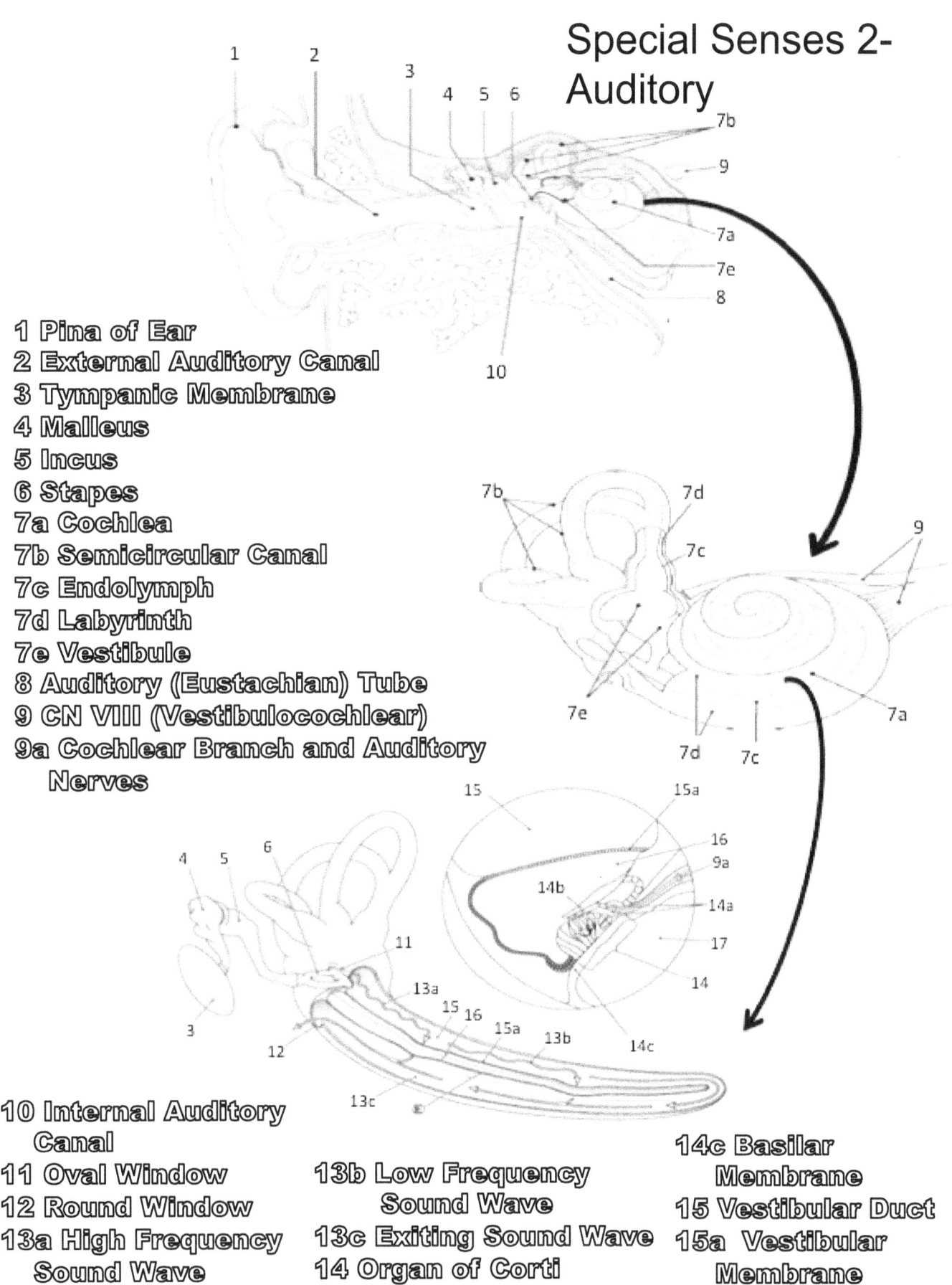

1 Pina of Ear
2 External Auditory Canal
3 Tympanic Membrane
4 Malleus
5 Incus
6 Stapes
7a Cochlea
7b Semicircular Canal
7c Endolymph
7d Labyrinth
7e Vestibule
8 Auditory (Eustachian) Tube
9 CN VIII (Vestibulocochlear)
9a Cochlear Branch and Auditory Nerves
10 Internal Auditory Canal
11 Oval Window
12 Round Window
13a High Frequency Sound Wave
13b Low Frequency Sound Wave
13c Exiting Sound Wave
14 Organ of Corti
14a Hair Cells
14b Tectorial Membrane
14c Basilar Membrane
15 Vestibular Duct
15a Vestibular Membrane
16 Cochlear Duct
17 Tympanic Duct

Special Senses 3- Vestibular

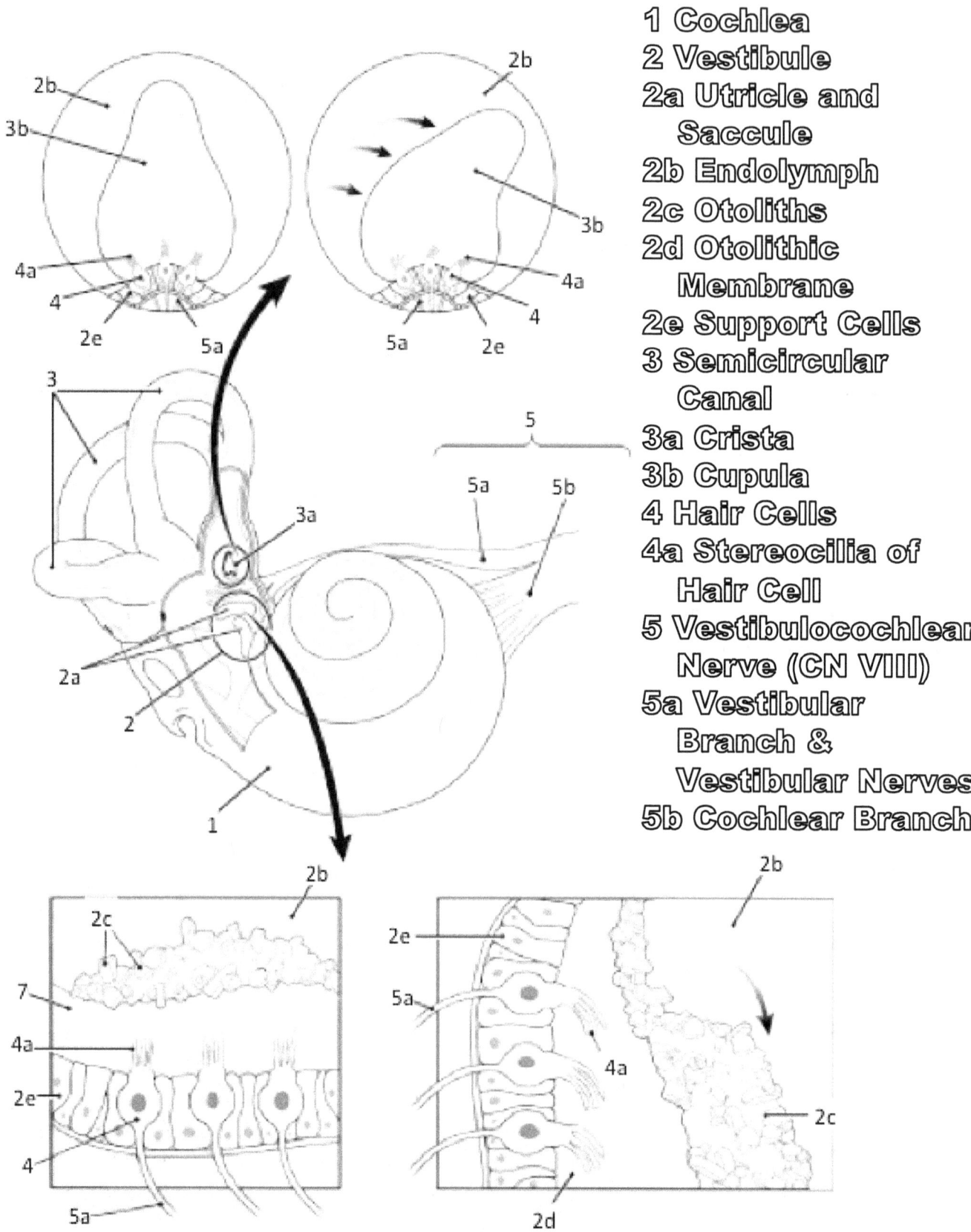

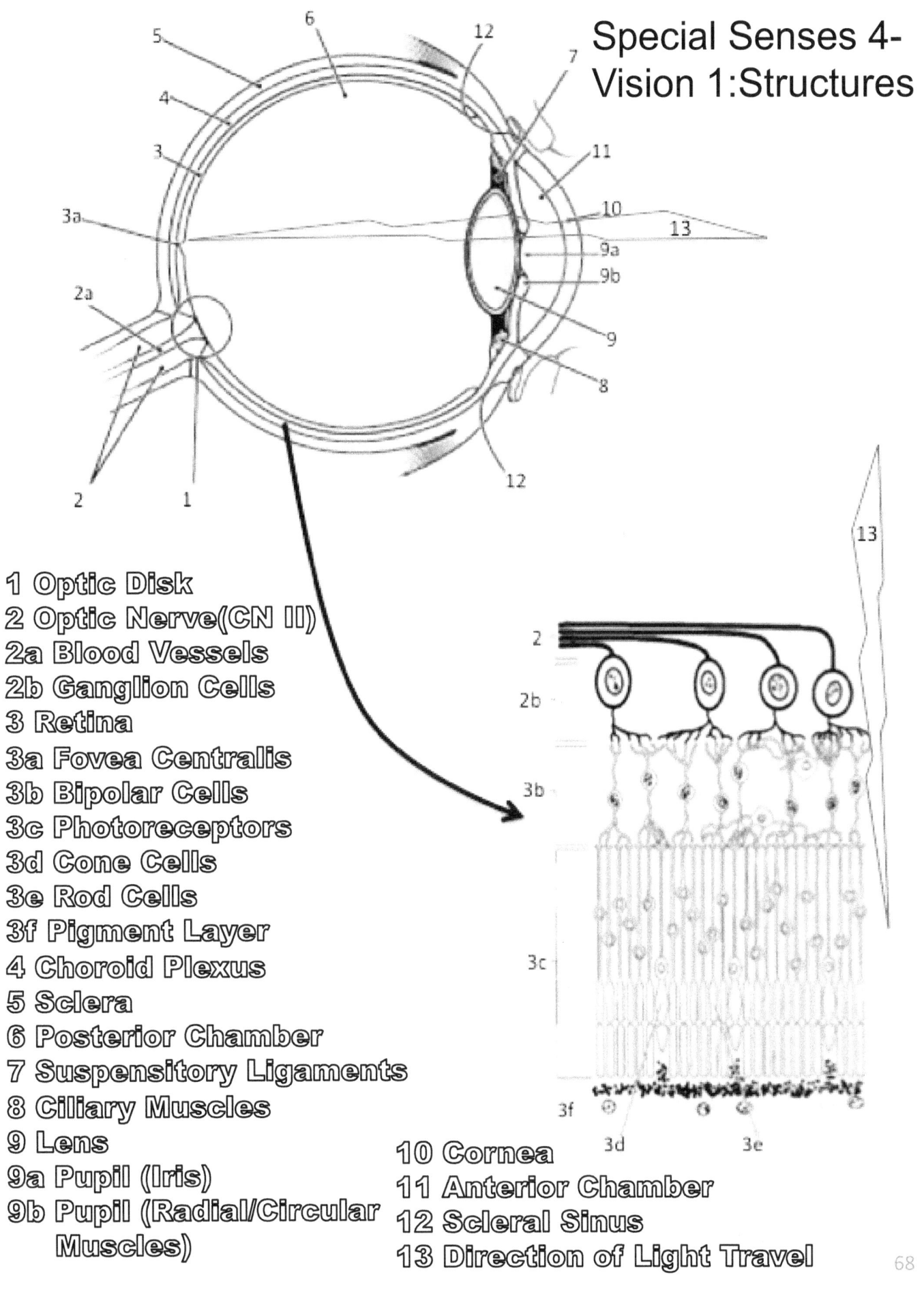

Special Senses 4 - Vision 1: Structures

1 Optic Disk
2 Optic Nerve (CN II)
2a Blood Vessels
2b Ganglion Cells
3 Retina
3a Fovea Centralis
3b Bipolar Cells
3c Photoreceptors
3d Cone Cells
3e Rod Cells
3f Pigment Layer
4 Choroid Plexus
5 Sclera
6 Posterior Chamber
7 Suspensitory Ligaments
8 Ciliary Muscles
9 Lens
9a Pupil (Iris)
9b Pupil (Radial/Circular Muscles)
10 Cornea
11 Anterior Chamber
12 Scleral Sinus
13 Direction of Light Travel

Special Senses 4- Vision 2

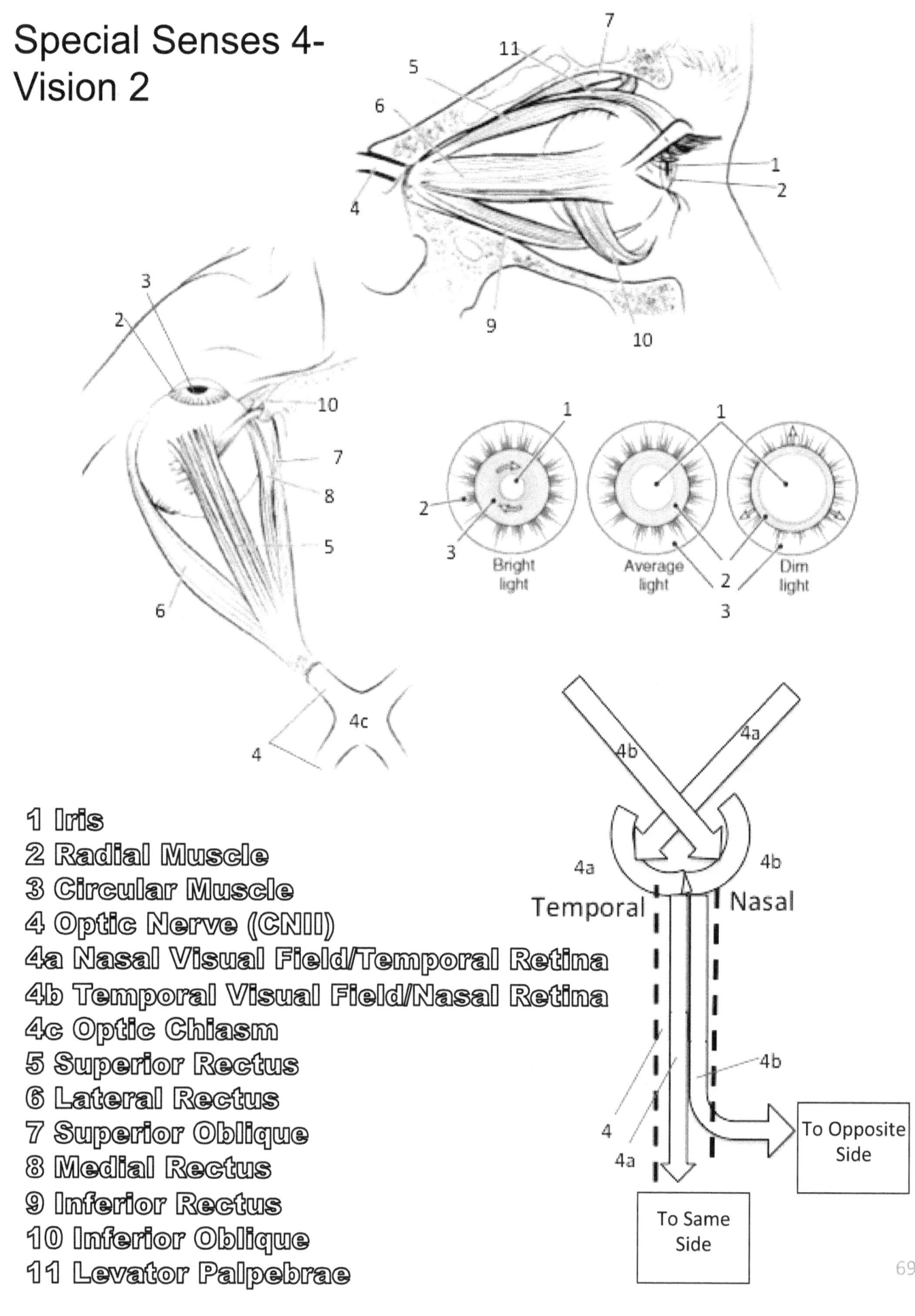

1 Iris
2 Radial Muscle
3 Circular Muscle
4 Optic Nerve (CNII)
4a Nasal Visual Field/Temporal Retina
4b Temporal Visual Field/Nasal Retina
4c Optic Chiasm
5 Superior Rectus
6 Lateral Rectus
7 Superior Oblique
8 Medial Rectus
9 Inferior Rectus
10 Inferior Oblique
11 Levator Palpebrae

Sensation 1- Peripheral Sensory Receptors

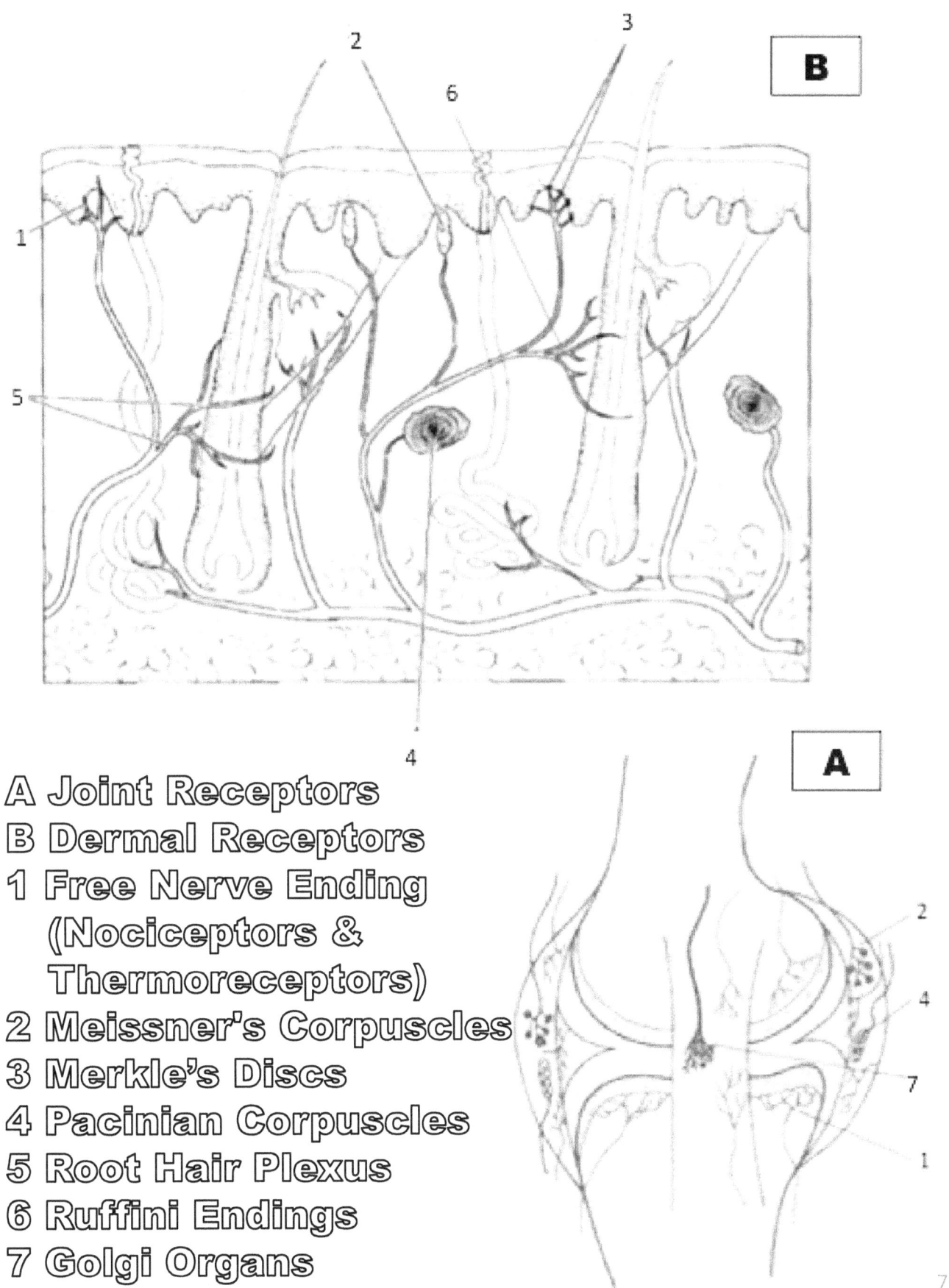

A Joint Receptors
B Dermal Receptors
1 Free Nerve Ending (Nociceptors & Thermoreceptors)
2 Meissner's Corpuscles
3 Merkle's Discs
4 Pacinian Corpuscles
5 Root Hair Plexus
6 Ruffini Endings
7 Golgi Organs

Sensation 2 - Somatosensory Cortex Topographic Organization

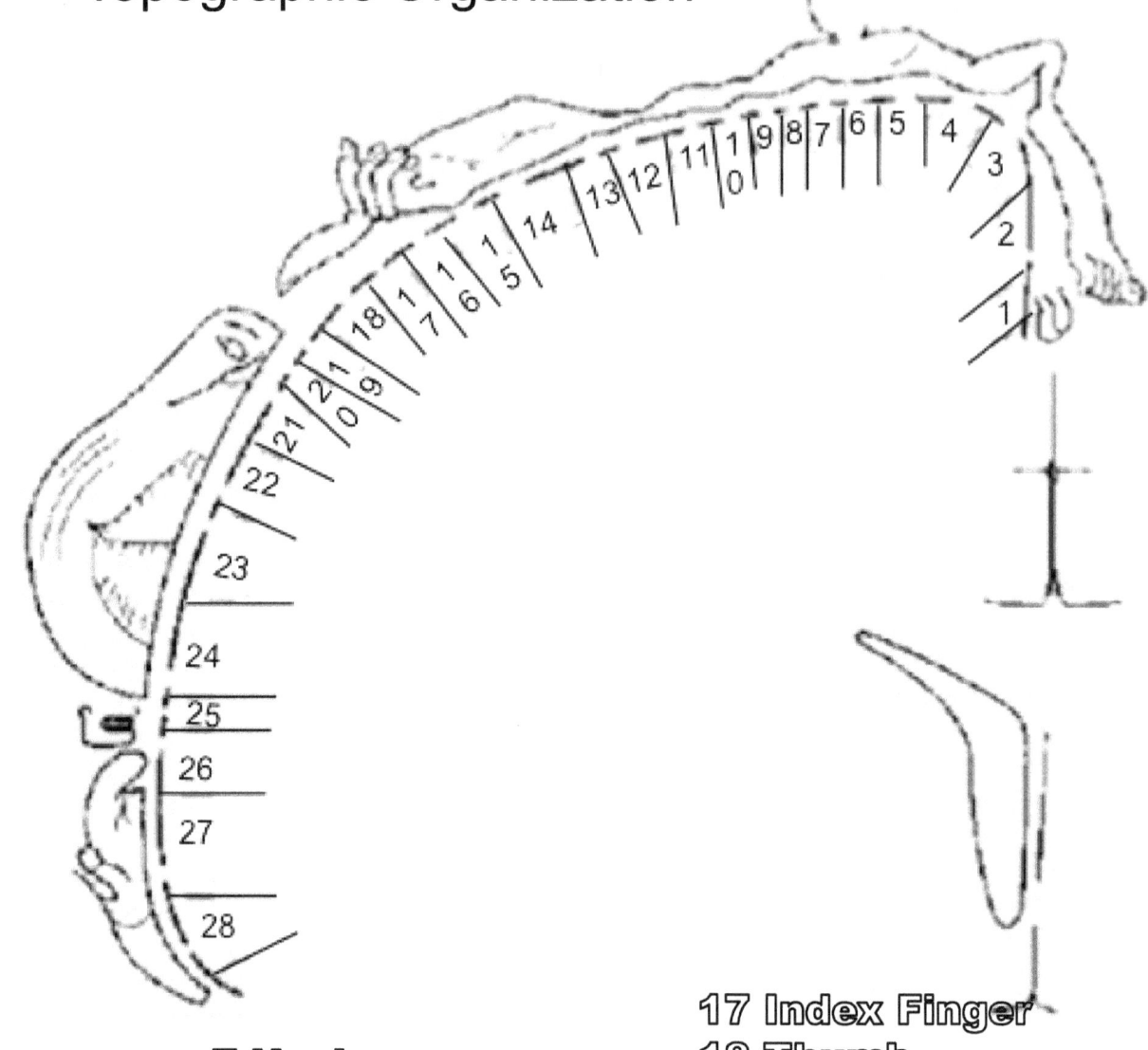

1 Genitals
2 Toes
3 Foot
4 Leg, Knee & Thigh
5 Hip
6 Trunk
7 Neck
8 Head
9 Shoulder
10 Arm
11 Elbow
12 Forearm
13 Wrist
14 Hand
15 Digiti Minimi & Ring Finger
16 Ring & Middle Finger
17 Index Finger
18 Thumb
19 Eye
20 Nose
21 Face
22 Upper Lips
23 Lips
24 Lower Lip
25 Tongue, Gums, Jaw
26 Tongue
27 Pharynx
28 Abdomen & Gut

Neuroendocrine System

Hormones, Glands and Tissues
involved with homeostasis

Overview

The system will utilize chemical messengers to transmit signals for changes that are necessary. In this case, the chemicals being released are called **CYTOKINES** and there are a few distinct categories of cytokines. If the chemical signals are neurological in origin and action, we call them **NEUROTRANSMITTER**. If the chemical signal is something that instead interacts with peripheral tissues then, then we call them **HORMONES**. Because chemical signals send messages of what is occurring and how changes in **HOMEOSTASIS** are impacting the tissues, **ALL TISSUES OF THE BODY CAN BE CONSIDERED TO BE HORMONE PRODUCING TISSUES** and not just the classically described "endocrine" glands. It is these peripheral signals (hormones) that we spend our time discussing. Using these chemical signals, the system is able to:

1) serve as a **MEANS FOR CELLS TO COMMUNICATE** between each other using chemical signals (aka hormones);

2) **REGULATE A NUMBER OF METABOLIC PROCESSES** throughout the body;

3) have distinct **ACTIONS THAT ARE PRECISE AND ONLY AFFECT SPECIFIC TARGET CELLS**. Where a specific receptor (membrane protein) is present for that hormone;

4) **COORDINATE A FAST AND SLOW RESPONSE TO STRESS** that helps **MAINTAIN HOMEOSTASIS** within the body.

The Neuroendocrine system is the system of interaction between the excitable nervous system tissues and the excretory tissues of the endocrine and exocrine glands. The interaction occurs at the junction between the pituitary and hypothalamus of the cerebral cortex. In this region there is an integration of excitatory and inhibitory signals that are feeding back onto the region both from nervous input, as well as from the circulatory system. Regardless of the method of feedback, the hypothalamus and pituitary are responsible for maintaining the homeostatic changes that occur due to stress or non-stress on the body by regulating the genetic signalling that occur within specific cells of distinct tissues (referred to as target tissues) of each of the various systems and the structures within those systems of the body. The organs and tissues within the system are widely distributed through the body and are classified by the type and method of regulation of secretion.

Neuroendocrine 1 - Tissue Overview

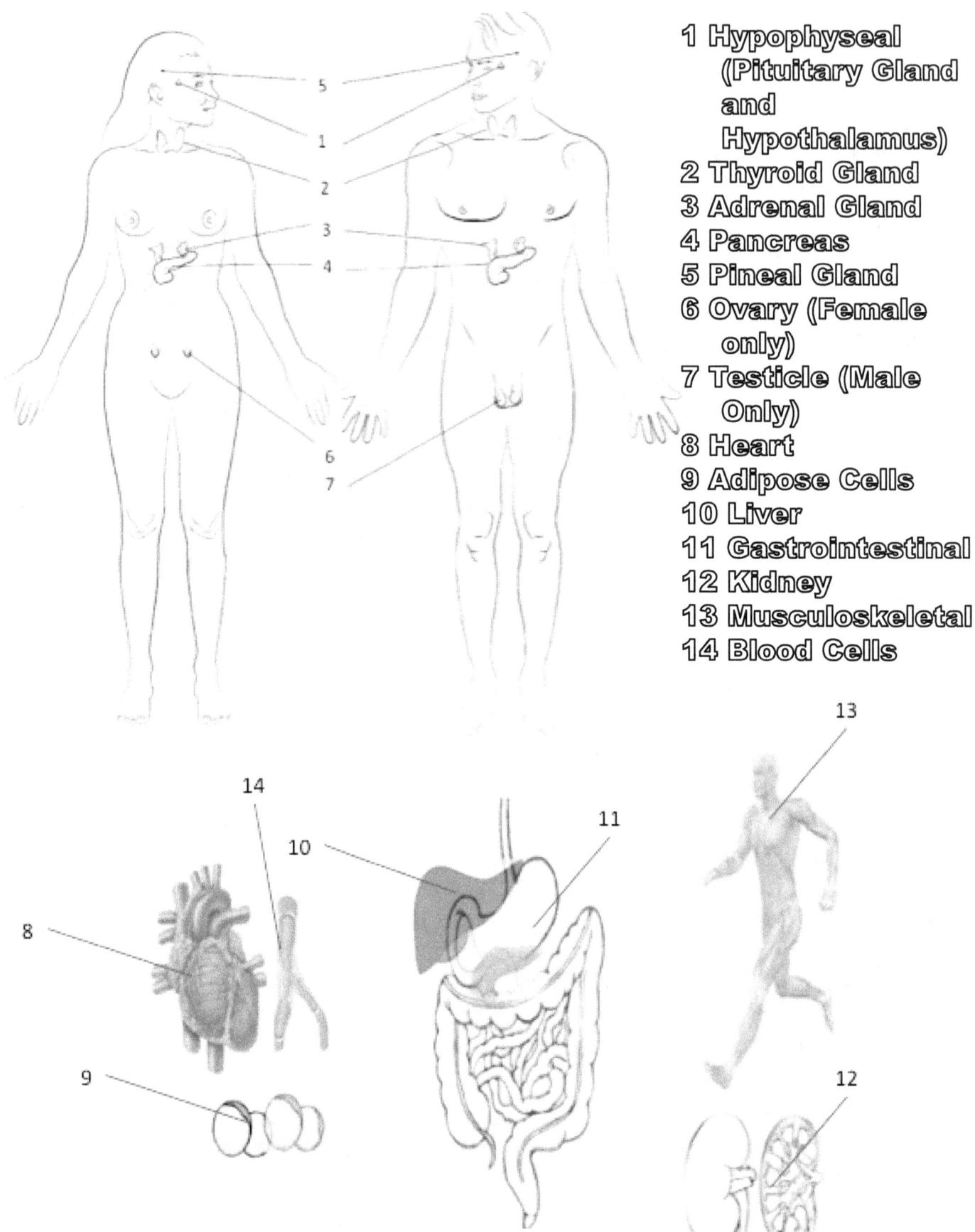

1. Hypophyseal (Pituitary Gland and Hypothalamus)
2. Thyroid Gland
3. Adrenal Gland
4. Pancreas
5. Pineal Gland
6. Ovary (Female only)
7. Testicle (Male Only)
8. Heart
9. Adipose Cells
10. Liver
11. Gastrointestinal
12. Kidney
13. Musculoskeletal
14. Blood Cells

Neuroendocrine 2- Pituitary

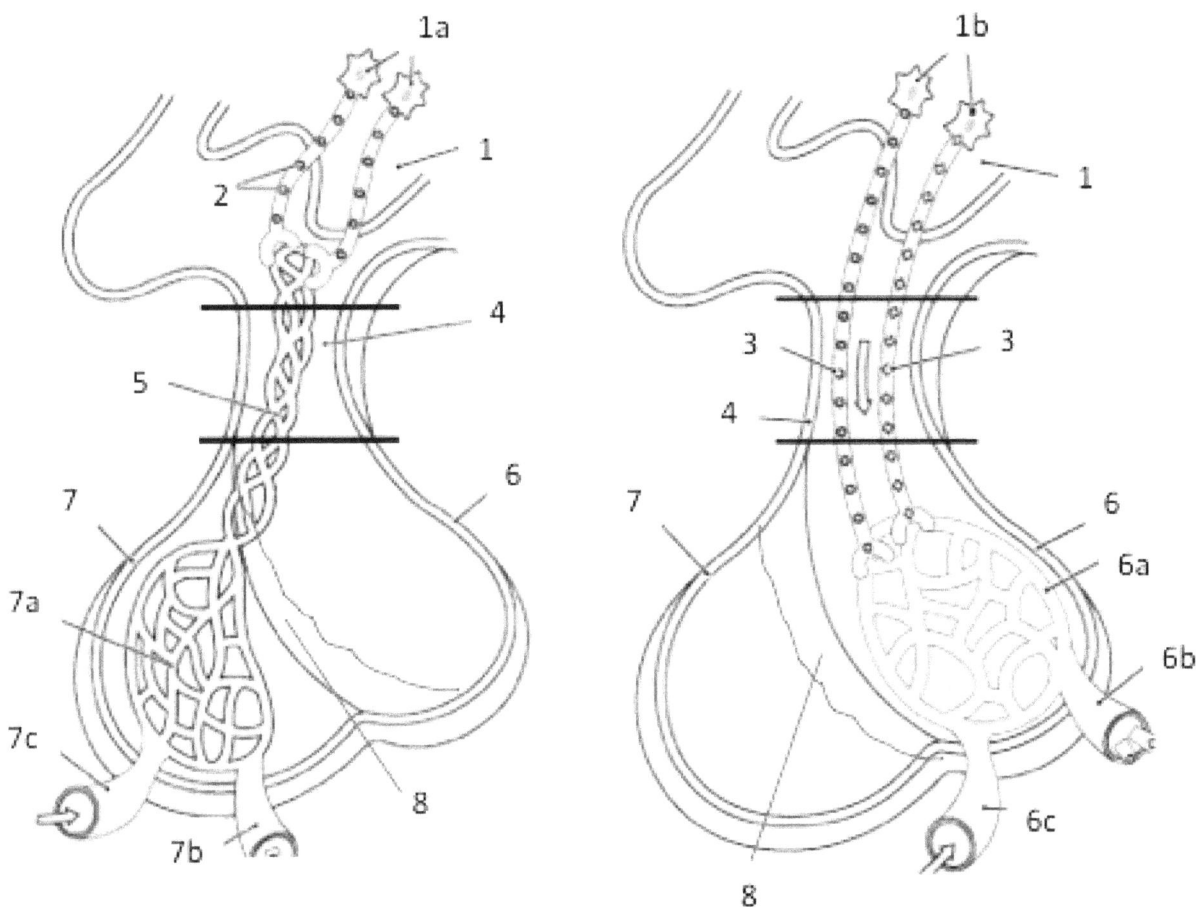

1=Hypothalamus
1a=Trophic Regulatory Neurons
1b=Oxytocin & Vasopressin Neurons
2=Axon Projections to Hypophyseal Stalk
3=Axon Projections to Posterior Pituitary
4=Infundibulum
5= Hypophyseal Portal
6=Posterior Pituitary Gland & Pars Nervosa
6a=Posterior Capillary Bed
6b=Efferent Arteriole
6c=Afferent Arteriole
7=Anterior Pituitary Gland & Pars Distalis
7a=Capillary Bed and Tropic/Trophic Cells
7b=Efferent Arteriole
7c=Afferent Arteriole
8=Pars Intermedia

Neuroendocrine 3 - Thyroid & Parathyroid

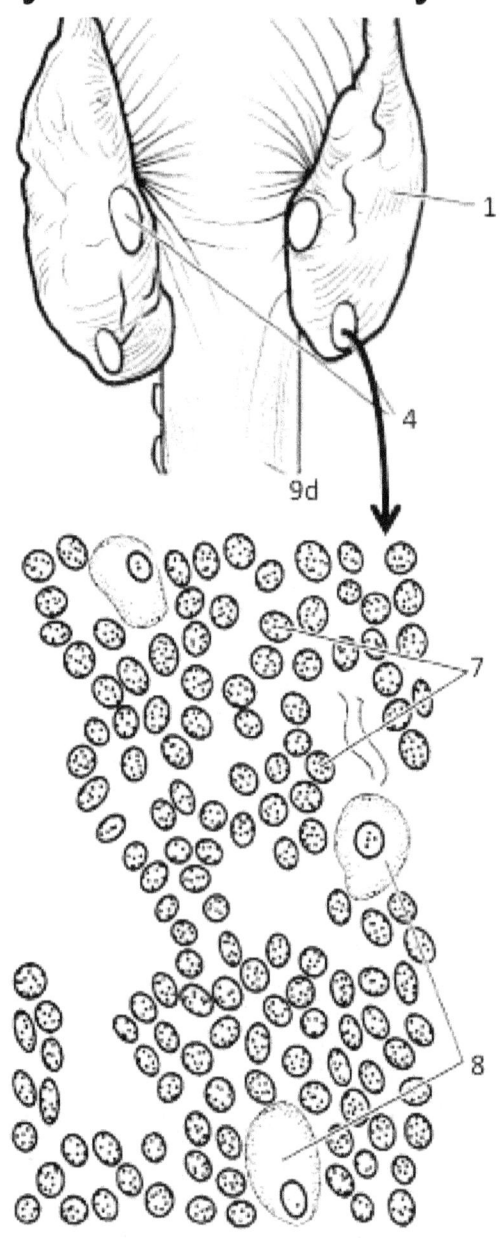
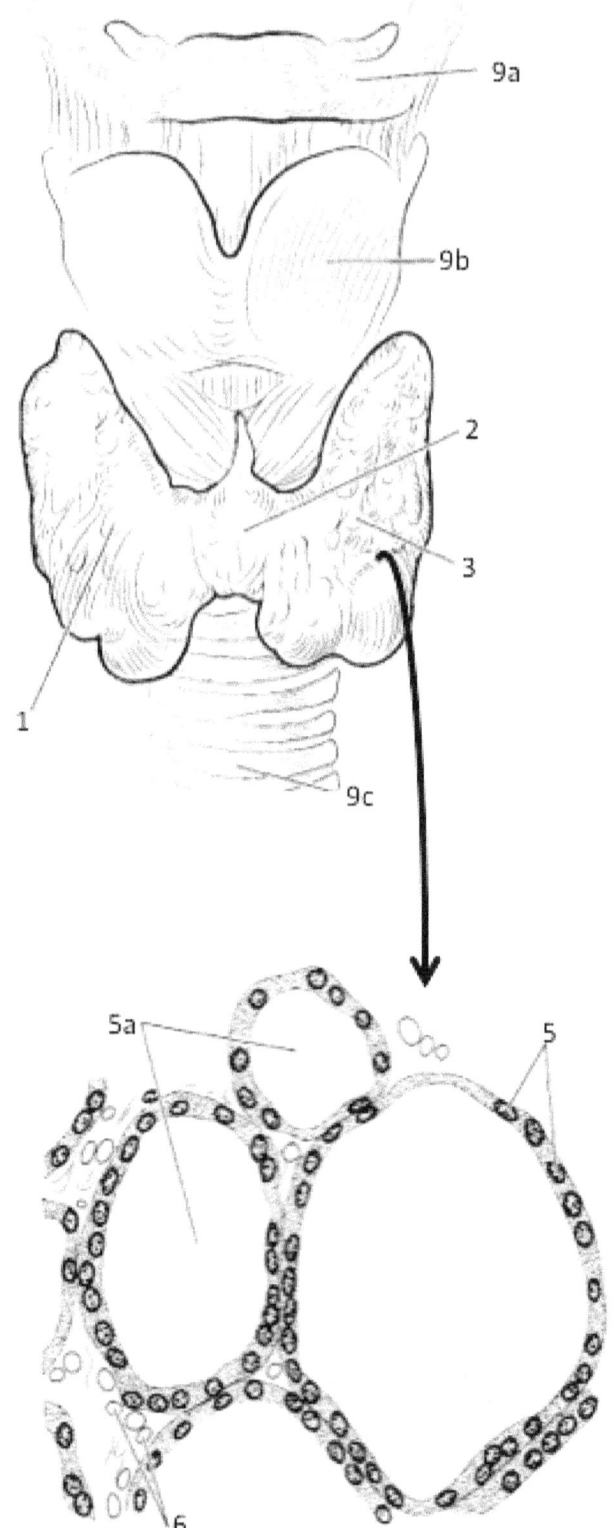

1 Thyroid Gland
2 Isthmus of Thyroid
3 Lobe of Thyroid
4 Parathyroid Gland
5 Follicular Cells (Produce T_3 & T_4)
5a Follicles of Thyroid
6 Parafollicular Cells (Produce Calcitonin)
7 Principle (Chief) Cells (Produce Parathyroid Hormone)
8 Oxyphilic Cells of Parathyroid
9a Hyoid Bone
9b Thyroid Cartilage
9c Trachea
9d Esophagus

Neuroendocrine 4- Adrenal Gland

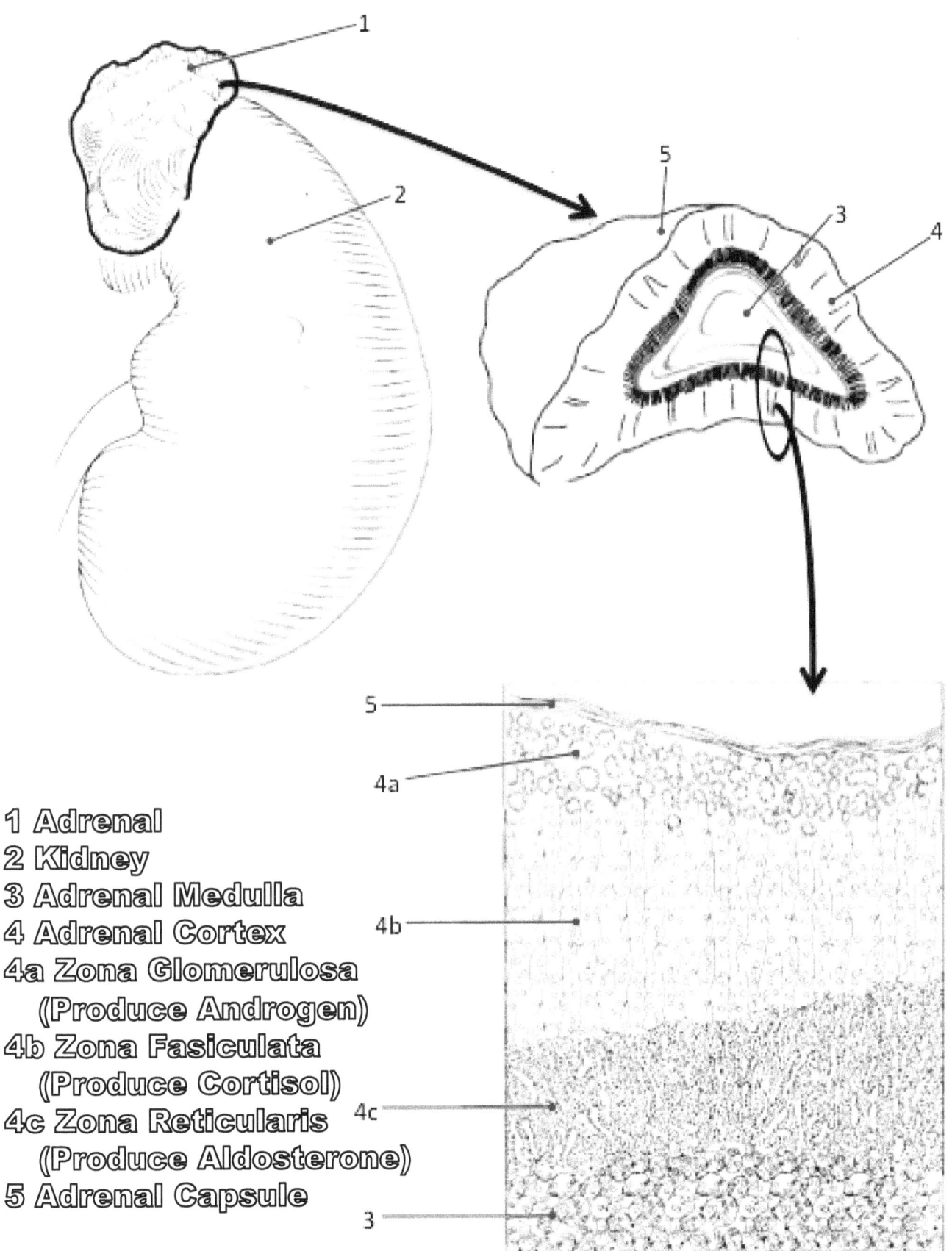

1 Adrenal
2 Kidney
3 Adrenal Medulla
4 Adrenal Cortex
4a Zona Glomerulosa
   (Produce Androgen)
4b Zona Fasiculata
   (Produce Cortisol)
4c Zona Reticularis
   (Produce Aldosterone)
5 Adrenal Capsule

Neuroendocrine 5- Pancreas

1 Tail of Pancreas
2 Body of Pancreas
3 Head of Pancreas
4 Pancreatic Duct
5 Arteriole
6 Islet of Langerhans
6a Alpha Cells (Glucagon producers)
6b Beta Cells (Insulin Producers)
7 Acinar Cells

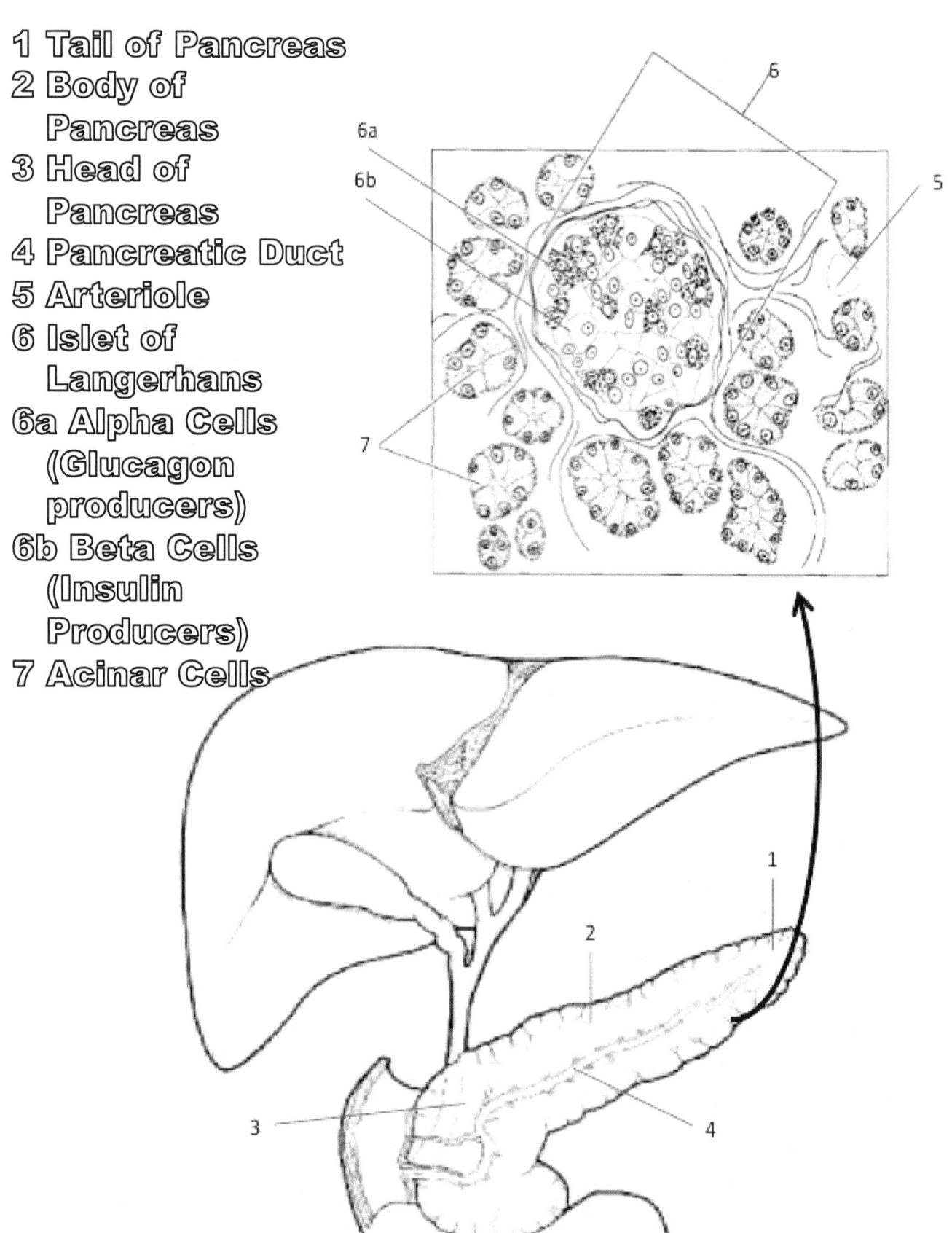

Neuroendocrine 6 - Gonads

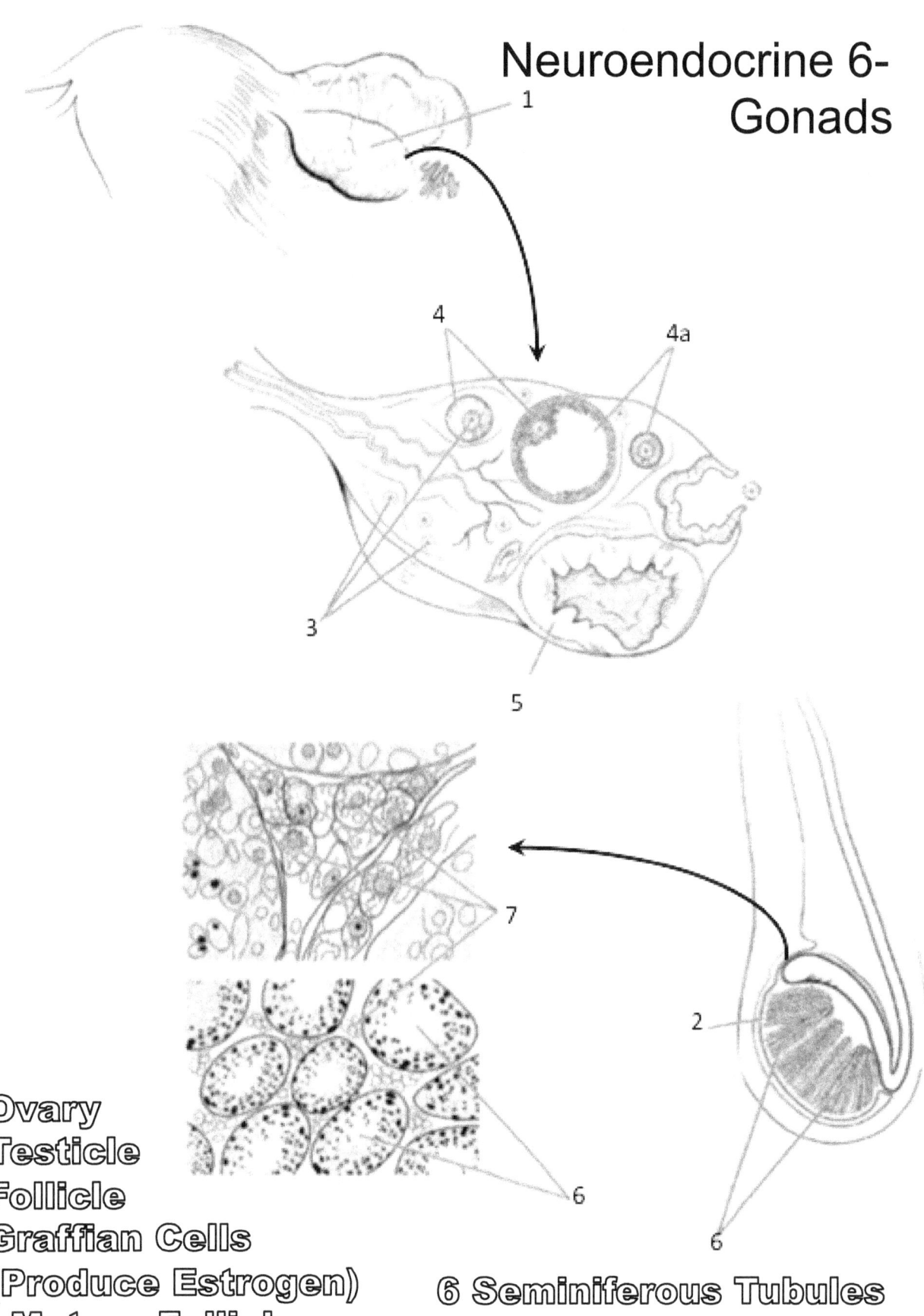

1 Ovary
2 Testicle
3 Follicle
4 Graffian Cells
   (Produce Estrogen)
4a Mature Follicle
5 Corpus Albicans
   (Produce Progesterone)
6 Seminiferous Tubules
7 Sertoli Cells
   (Produce Testosterone)

Cardiorespiratory System

Heart, Blood, Blood Vessels, Lungs and Gas Exchanges

Cardiovascular Overview

The cardiorespiratory system is comprised of two linked systems for the body, the CARDIOVASCULAR and the RESPIRATORY systems. The systems are integrated in such a way that is almost impossible to discuss the physiological role of one system for the body without also including details and functions of the other system. The two systems function to ensure that GAS EXCHANGE occurs between the environment and body and between tissues of the body. Ensuring that AEROBIC METABOLISM can be completed at tissues and cells of the body.

Actions in the cardiovascular system will also ensure that ACID/BASE balance of the tissues is constant and that the body is able to maintain a constant core body temperature, THERMOREGULATION. The cardiovascular system performs this function through a closed network of tubes, ARTERIES and VEINS, and thin walled or perforated tubes CAPILLARIES the move BLOOD through the body through positive pressures in the arteries and actions at the HEART through contraction (SYSTOLE) of the VENTRICLES. The overall performance of the heart, or "fitness" of the heart, is expressed through the inverse relationship between HEART RATE (number of contractions of the ventricle in a minute) and STORKE VOLUME (amount of blood moved with each contraction of the ventricle) that is based on the amount of blood returning to the heart (VENOUS RETURN) and the level of resistance to the movement of blood during systole (PERIPHERAL RESISTANCE). Some have mistakenly stated that changes are due to changes in a supposed "need for Oxygen" by the body. But there is a problem as there are no oxygen sensors that cause the system to change. Another misconception is that blood can be seen as being OXYGENATED and DEOXYGENATED, however that is not true as OXYGEN IS ALWAYS PRESENT in the blood. The presence of oxygen is the reason that blood is ALWAYS RED IN BOTH ARTERIES AND VEINS. More oxygen present more red, less oxygen less red.

Key Terms for the System

In examining the cardiorespiratory systems is important to understand the following key terms

BULK FLOW: the concept that all materials within a solution will move a single unit in the direction of high pressure to low pressure

AFFINITY: is the concept related to the likelihood that select molecules are to interact, bone and remain bound each other relative to other molecules

VENOUS RETURN: amount of blood (by volume) that enters the right atrium for each heart beat

CARDIAC OUTPUT: amount of blood (by volume) that is ejected by the left ventricle per minute

CARDIAC SHIFT (CARDIAC DRIFT): Indicates that the rate of the heart contraction will change based on the reduction of venous return so as to keep cardiac output stable

COMPLIANCE: the ability for the artery and arteriole, or bronchi and bronchiole, to change the internal diameter to allow for a change in the volume of blood, or air, to move through the lumen

ANGIOGENESIS: the process of developing new capillary networks, arterioles and venuoles within tissues of the body

FRANK-STARLING LAW: Details how the stretch of the ventricle by the volume of fill and lower peripheral resistance can lead to a greater passive force of contraction, larger stroke volume and lower heart rate

STARLING'S LAW: Details how the difference between hydrostatic/colloid pressure and osmotic pressure across a membrane will determine the direction of bulk flow materials

Cardiorespiratory-Overview

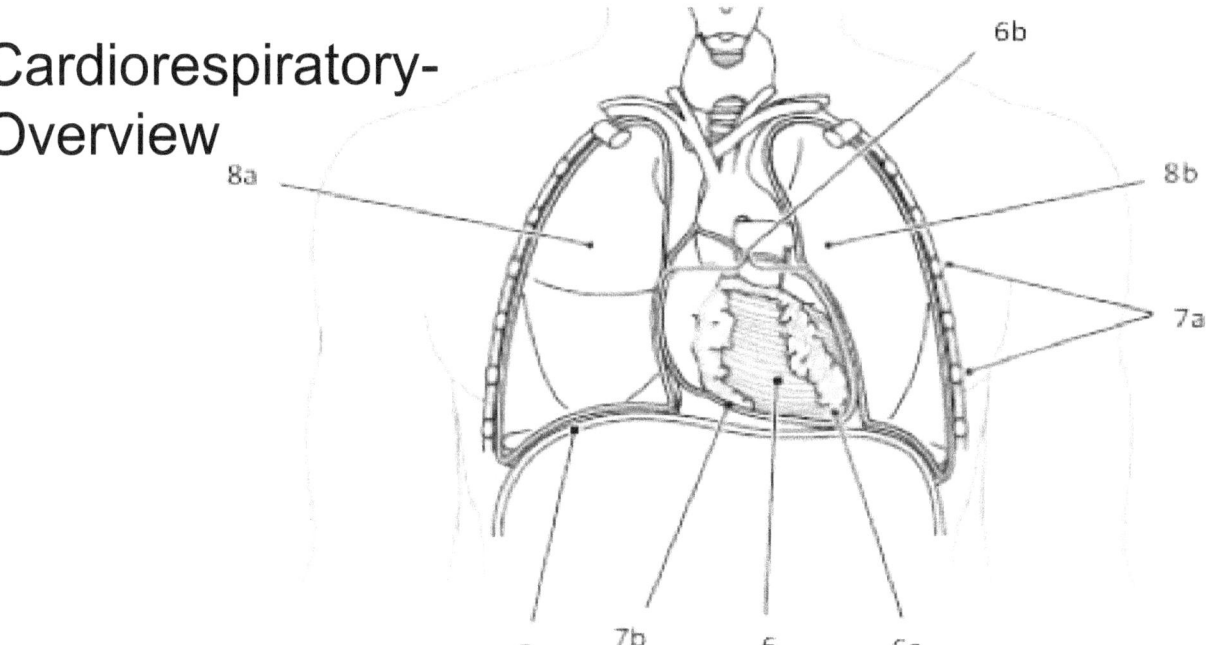

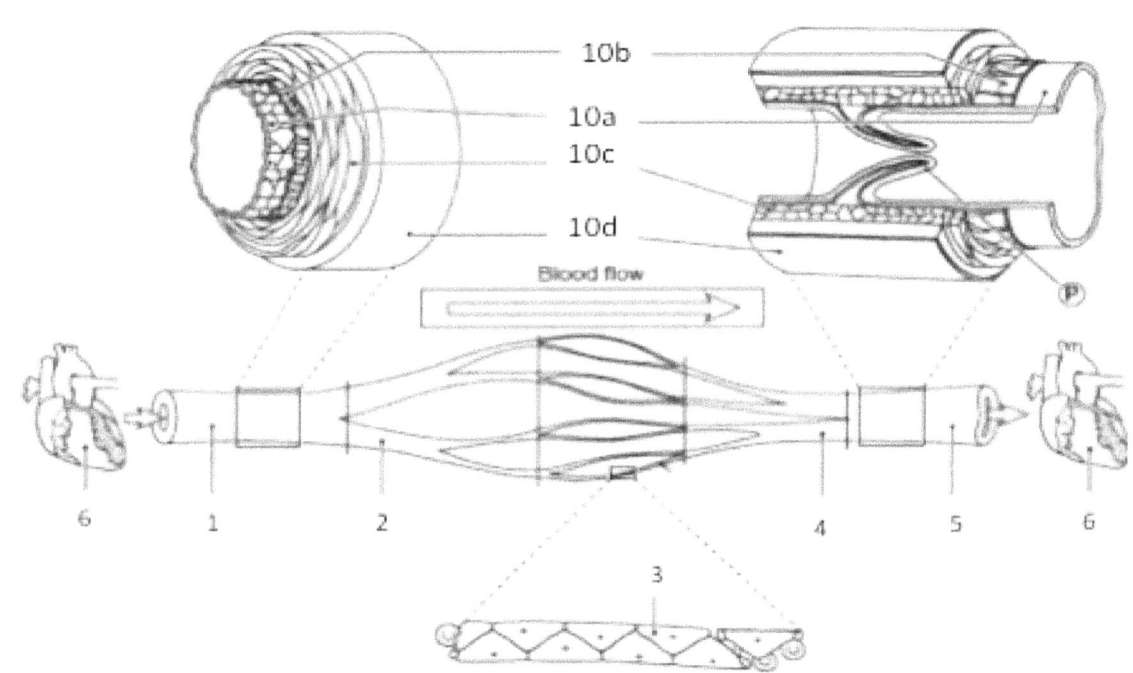

1 Artery
2 Arteriole
3 Capillary
4 Venuole
5 Vein
6 Heart
6a Ventricle
6b Atria

7a Thoracic Cage/Pleural Lining
7b Pericardium/Mediastinum
8a Right Lung
8b Left Lung
9 Diaphragm
10a Tunica Intima
10b Elastic Layer
10c Tunica Media (Musclaris)
10d Tunica Adventitia

Cardiovascular 1 - Vessel Pathway

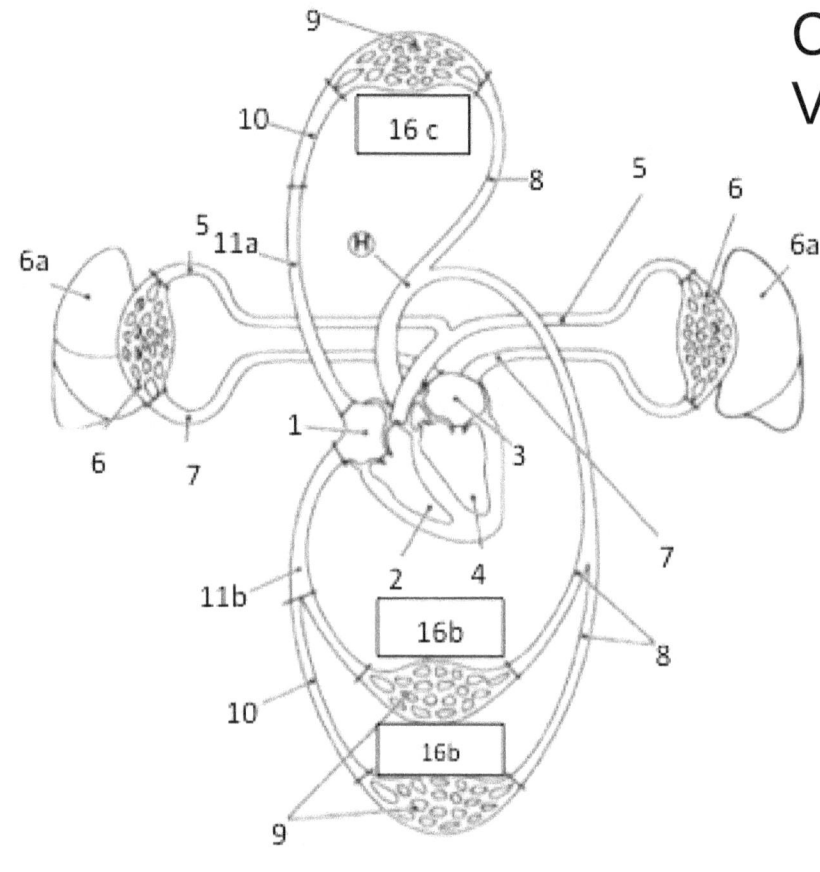

1. Right Atria
2. Right Ventricle
3. Left Atria
4. Left Ventricle
5. Pulmonary Artery
6. Pulmonary Capillary
6a. Lungs
7. Pulmonary Vein
8. Systemic Artery
8a. Arteriole
9. Systemic Capillary
10. Systemic Veins
10a. Venuole
11a. Superior Vena Cava
11b. Inferior Vena Cava
12. Anastomosis
12a. Circular Muscle/Sphincter
13. Red Blood Cell (Erythrocyte)
14. Oxygen
15. Carbon Dioxide
16. Tissues of Body
16a. Cells of Tissues
16b. Organs and Lower Extremity
16c. Head, Brain and Upper Extremity

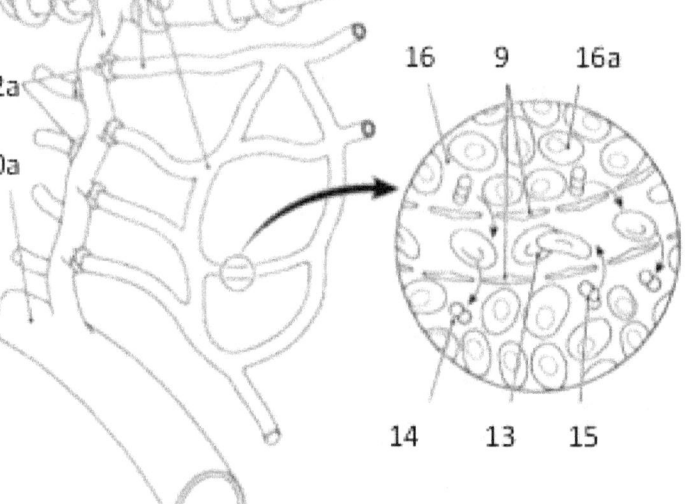

Cardiovascular 2- Blood

1 Whole Blood
2 Plasma
2a Water
2b Dissolved Substances
2c Proteins
3 Formed Elements/ Hematocrit
4 Red Blood Cells (Erythrocytes)
4a Hemoglobin
5 Buffy Coat
5a Eosinophil
5b Basophil
5c Neutrophil
5d Monocyte
5e Platelets
5f Lymphocyte
5g Nucleus
5h Granules

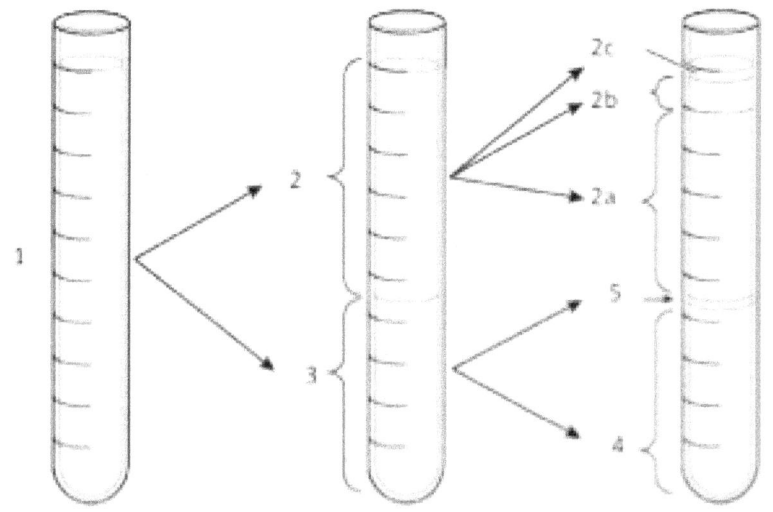
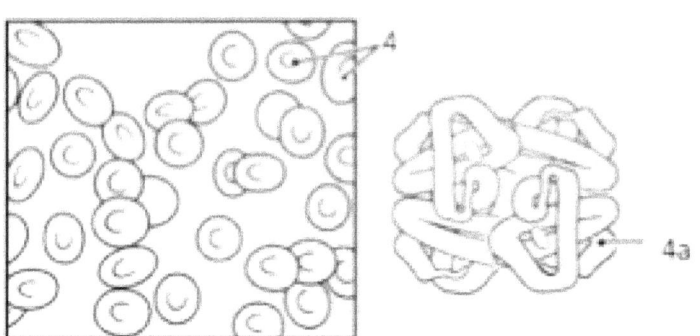
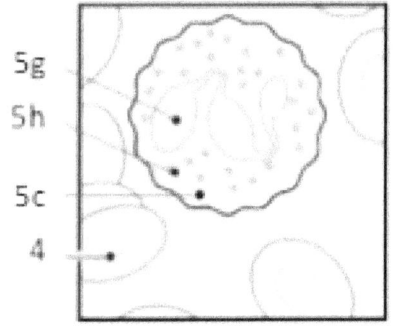
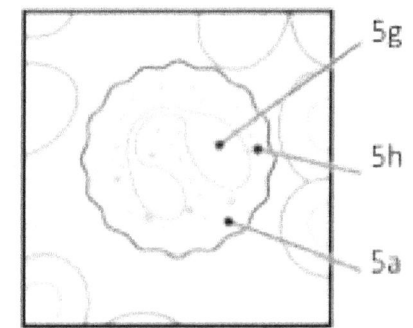
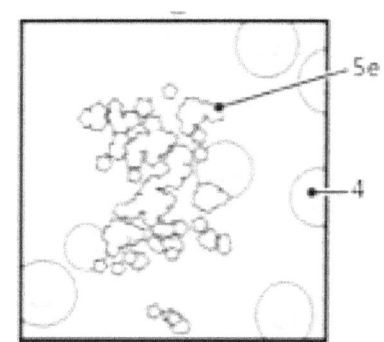
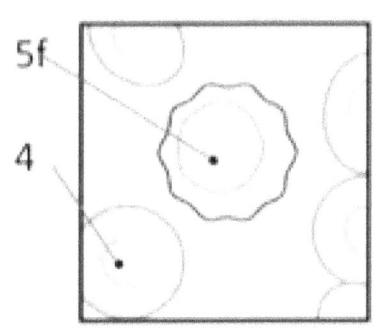
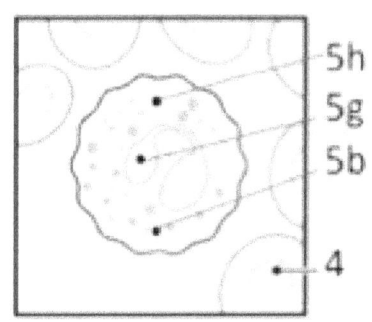
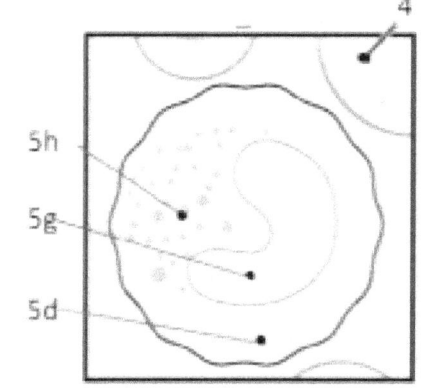

Cardiovascular 3- Heart 1-External Anatomy

Anterior

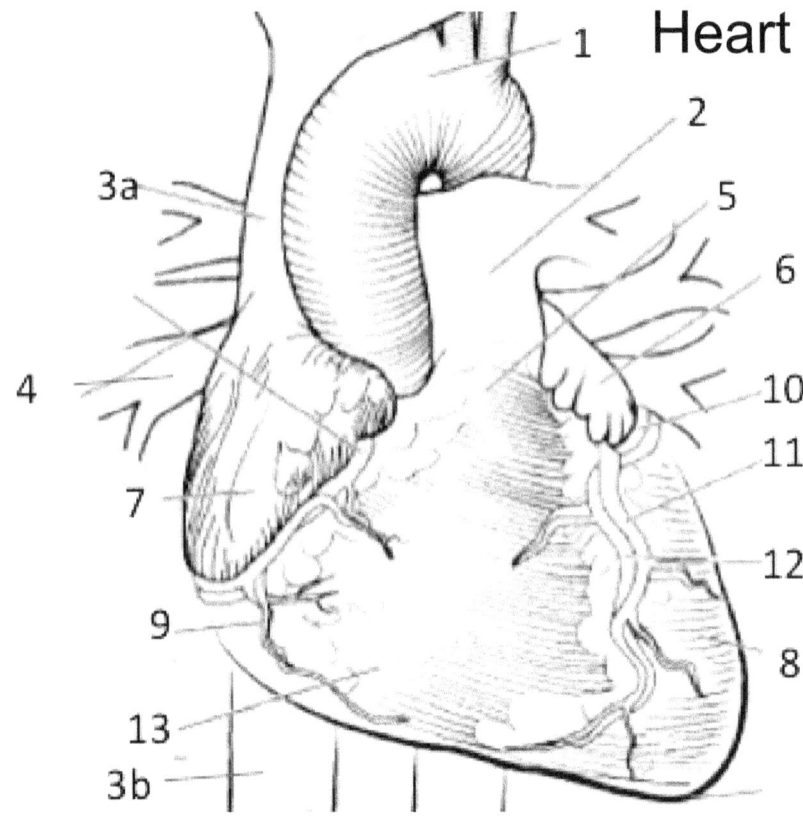

1 Aorta
2 Pulmonary Trunk and Arteries
3a Superior Vena Cava
3b Inferior Vena Cava
4 Pulmonary Veins
5 Septal Divide
6 Auricle of Left Atria
7 Auricle of Right Atria
7a Right Atria
8 Left Ventricle
9 Marginal Artery & Small Cardiac Vein
10 Left Atria
11 Great Cardiac Vein
12 Left Anterior Descending Interventricular Anterior
13 Right Ventricle
14 Posterior Interventricular Artery
15 Middle Cardiac Vein
16 Cardiac Sinus

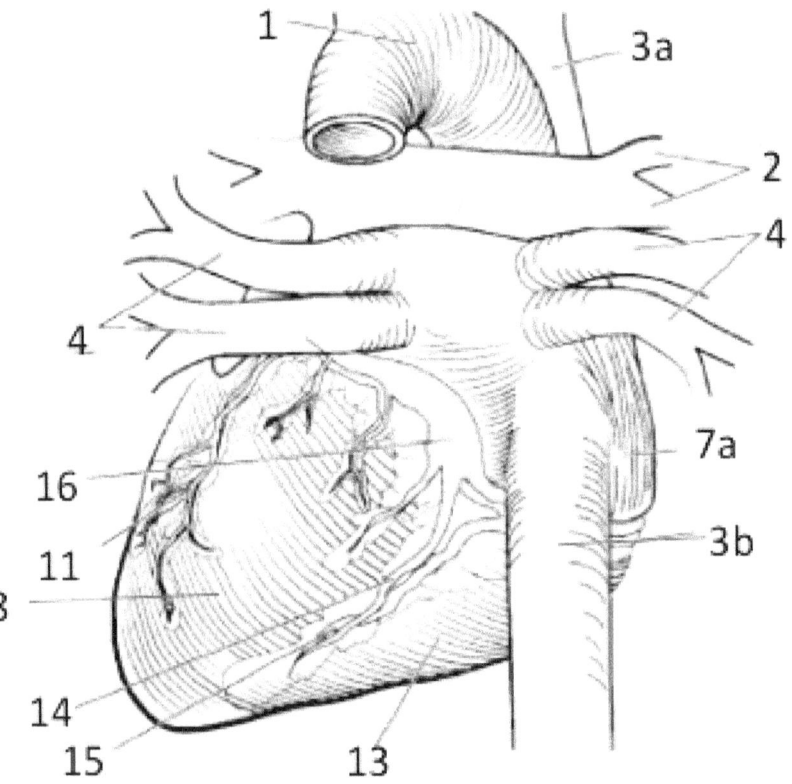

Posterior

Heart Properties

The heart is unique to the other organs of the body as the heart functions based on having an AUTOREGULATORY and AUOTRHYTMICITY behavior. That is the heart functions independent of any other regulation signals, and can even continue to function following death of the other organs of the body. These behaviors are established by specialized tissue within the myocardium that is responsible for the transmission of the electrical signal from the SINOATRIAL NODE (SA) to the ATRIOVENTRICULAR NODE (AV) and then through the ventricle of the heart.

Contraction of the heart is biphasic, SYSTOLE (contraction) and DIASTOLE (relaxation), within a cycle of activation that is further split to an ATRIA CONTRACTION CYCLE and a VENTRICLE CONTRACTION CYCLE. The heart never completely contracts (both atria and ventricle) because of its function of being a hydraulic pump that needs to move fluid somewhere. Rate of contraction is defined two ways: INTRINSIC, based on the set rate at which the SA node functions and threshold will spread through the atria to the atria-ventricular nodes and then through the conductive pathway of the ventricle, or RHYTHMIC, rate of contraction based on the stretch and contraction cycle that is generated through the FRANK-STARLING'S LAW that changes the intrinsic rate. Frank-Starling Law tells us that the rhythmic rates are altered by the level of venous return and the relationship between contraction force and total peripheral resistance (due to compliance of the arteries) as this will change the stroke volume and thus the heart rate inversely to the change in stroke volume.

Cardiovascular 3- Heart 2-Internal Anatomy

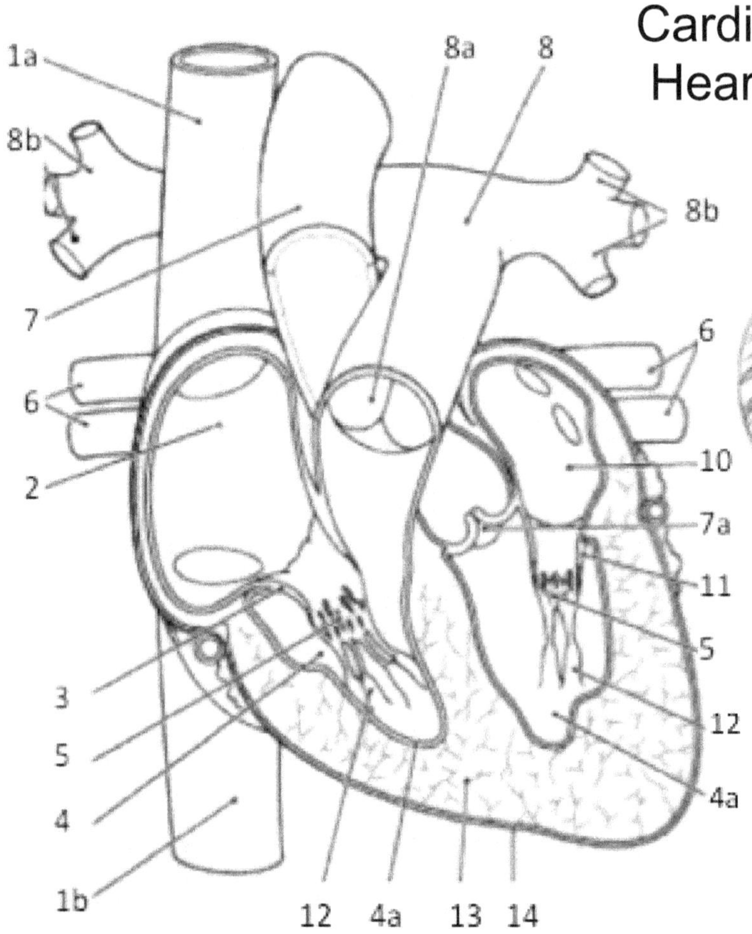

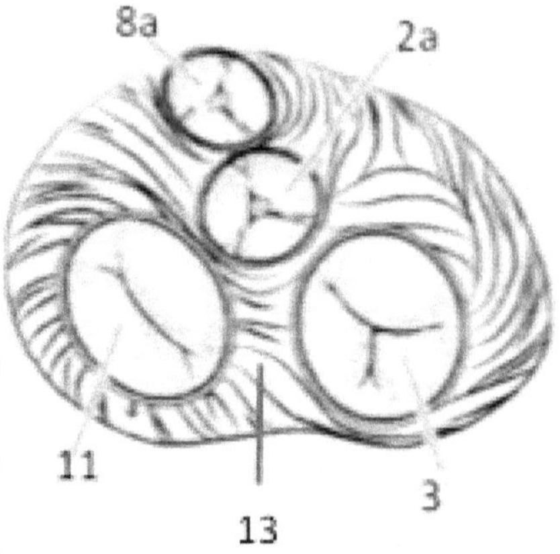

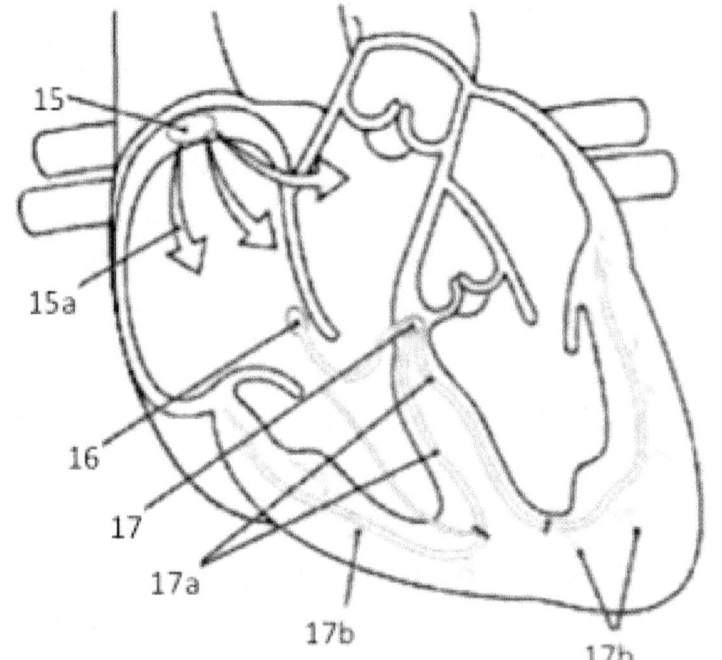

1a Superior Vena Cava
1b Inferior Vena Cava
2 Right Atria
3 Tricuspid Valve
4 Endocardium
4a Trabecular Carnae
5 Chordae Tendineae
6 Pulmonary Veins
7 Aorta
7a Aortic Valve
8 Pulmonary Trunk
8a Pulmonary Valve
9 Pulmonary Arteries
10 Left Atria
11 Bicuspid (Mitral) Valve
12 Papillary Muscle
13 Myocardium
14 Epicardium
15 Sino-Atrial (SA) Node
15a Atrial Conduction
16 Atrial-Ventricular (AV) Node
17 Bundle of His
17a Bundle Branches
17b Perkinje Fibers

Cardiovascular 4- Heart Cycle

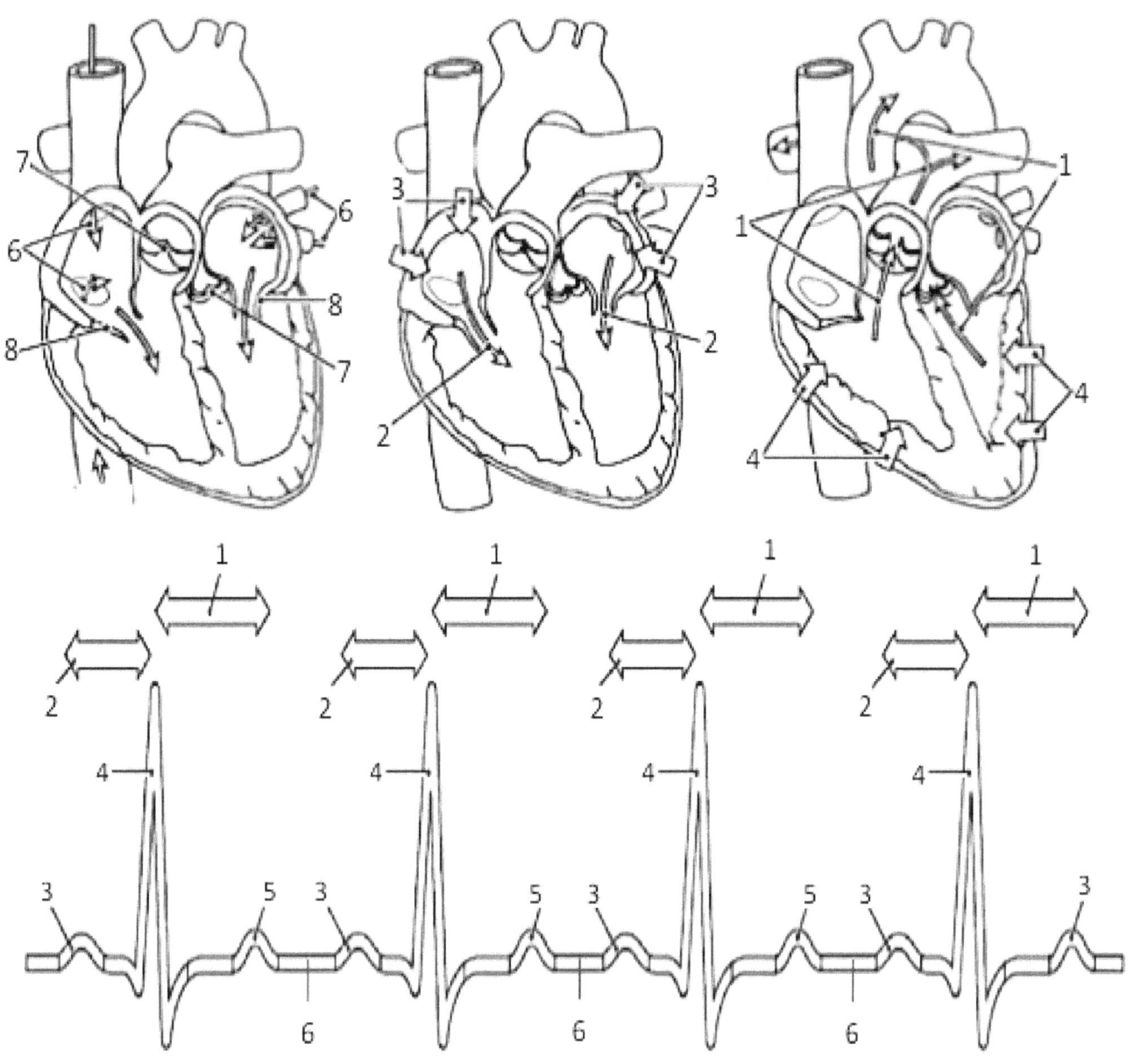

1 Ventricular Systole (Contraction)
2 Atrial Systole (Contraction)
3 P-Wave
4 QRS Complex
5 T-Wave
6 Heart Relaxed (Passive Fill)
7 Valve Closed
8 Valve Open

Cardiovascular 5a- Circulation of Blood through the Upper Extremities of the Body

Arteries

1 Aorta
1a Left Common Carotid Artery*
1b Left Subclavian Artery
1c Right Brachiocephalic Artery (Trunk)
2 Right Common Carotid Artery*
3 Right Subclavian Artery
4 Artery to Supraspinatus
5 Axillary Artery*
5a Lateral Thoracic Artery
6 Humeral Circumflex Artery
7 Brachial Artery*
8 Deep Brachial Artery
9 Nutrient Artery to Humerus
10 Radial Artery*
11 Ulnar Artery*
12a Deep Palmar Arch Artery
12b Superficial Palmar Arch Artery
12c Digital Artery
12d Radial Recurrent Artery

Veins

13 Cephalic Vein
14 Deltoid Vein
15 Brachial Vein
16 Basilic Vein
17 Axillary Vein
18 Antecubital Vein
19 Medial Median Antebrachial Vein
20 Radial Vein
21 Ulnar Vein
22 Middle Median Antebrachial Vein
23 Lateral Median Antebrachial Vein
24a Deep Venal Ach
24b Superficial Dorsal Venal Arch
24c Digital Veins

* Pulse Point Available

Cardiovascular 5b- Vessels of Upper Extremity

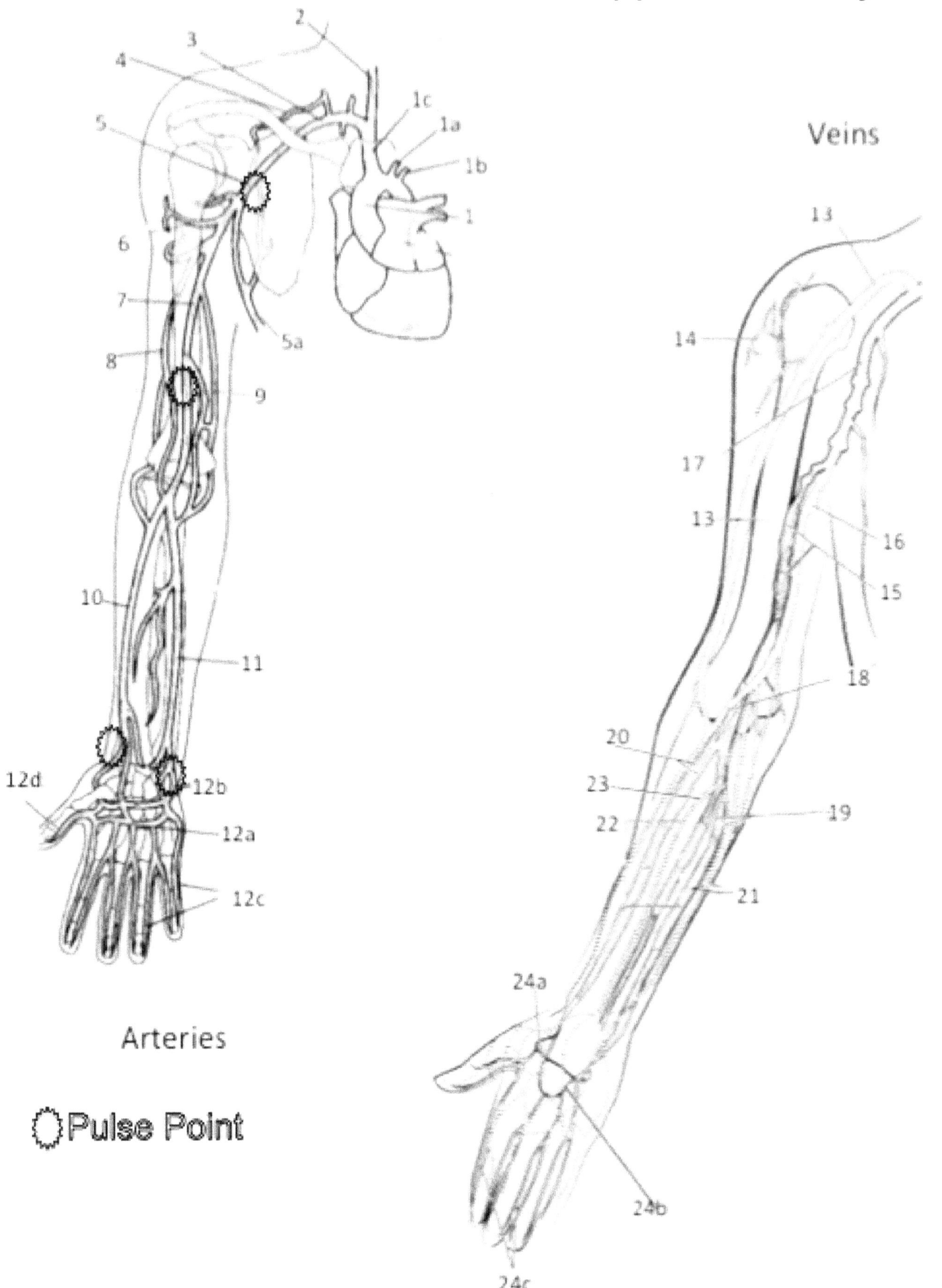

Cardiovascular 6a- Circulation of Blood through the Lower Extremities of the Body

Arteries

1 Common Iliac Artery
2 Internal Iliac Artery
3 External Iliac Artery
4 Femoral Artery*
5 Deep Femoral Artery
5a Femoral Circumflex Artery (to head of Femur)
6 Popliteal Artery* (on posterior side)
6a Geniculate Artery
7 Anterior Tibial Artery
8 Posterior Tibial Artery*
9 Peroneal Artery
10 Dorsalis Pedis Artery*
11 Dorsal Metatarsal Artery

Veins

12 Inferior Vena Cava
12a Common Iliac Vein
13 Internal Iliac Vein
14 External Iliac Vein
15 Femoral Vein
15a Deep Femoral Vein
16 Great Saphenous
17 Anterior Tibial
18 Popliteal
19 Smaller Saphenous
20 Posterior Tibial
21 Dorsal Venal Arch
21a Digital Veins
22 Plantar Venal Arch

* Pulse Point Available

Cardiovascular 6b- Vessels of Lower Extremity

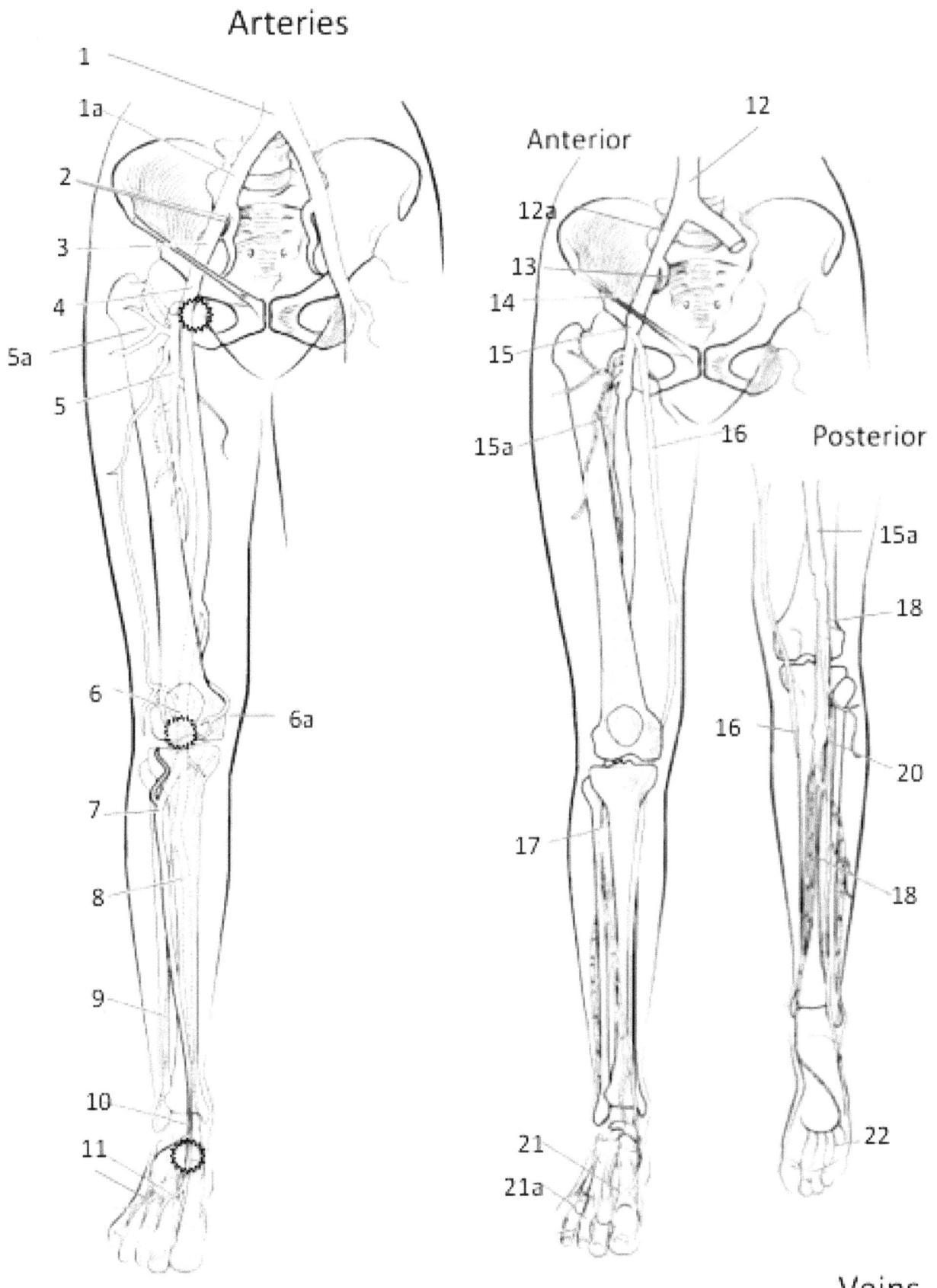

Cardiovascular 7a-Circulation of Blood through Thorax and Abdomen—Branches of Aorta and Vena Cava

Arteries
(Branches from Aorta)

1 Aorta
1a Aortic Arch
1b Coronary Arteries
1c Brachiocephalic Artery (Trunk)
2a Left Common Carotid Artery*
2b Right Common Carotid Artery*
2c Internal Carotid Artery
2d External Carotid*
3a Left Subclavian Artery
3b Right Subclavian Artery
4 Intercostal Arteries
5 Celiac Trunk
5a Left Gastric Artery
5b Splenic Artery
5c Hepatic Artery
6 Renal Artery
7 Gonadal Artery
8a Superior Mesenteric Artery
8b Inferior Mesenteric Artery
9 Common Iliac Artery
9a Internal Iliac Artery
9b External Iliac Artery

Veins
(Branches to Vena Cava)

10 Superior Vena Cava
11a Right Brachiocephalic
11b Left Brachiocephalic
12a Internal Jugular
12b External Jugular
13a Hemiazygos Vein
13b Azygos Vein
14 Intercostal Veins
15 Inferior Vena Cava
16 Hepatic Portal Vein
16a Hepatic Vein
17 Splenic Vein
18a Gastric Vein
18b Gastroepiploic Vein
19 Renal Vein
20 Gonadal Vein
21a Superior Mesenteric Vein
21b Inferior Mesenteric Vein
21c Colonic Vein
22 Common Iliac Vein
22a External Iliac Vein

*Pulse Point Available

Cardiovascular 7b- Vessels of Thorax and Abdomen

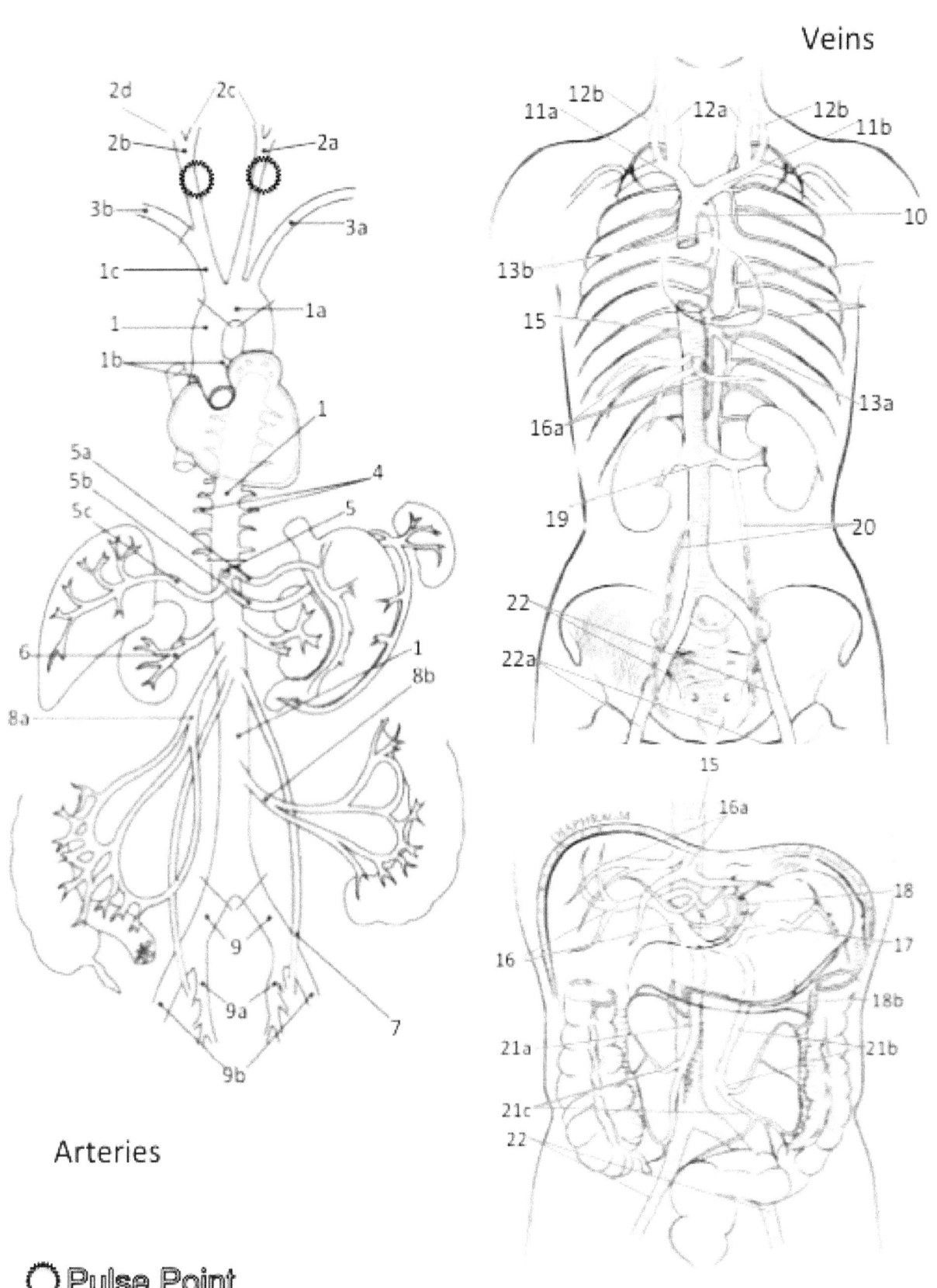

Arteries

○ Pulse Point

Cardiovascular 8a- Circulation through the Cranium and Cerebral Cortex

Arteries

1 Brachiocephalic Trunk (Artery)
1a Subclavian Artery
2 Common Carotid Artery*
2a External Carotid Artery*
2b Internal Carotid Artery
2c Thyroid Artery
3a Right Vertebral Artery
3b Left Vertebral Artery
4 Facial Artery
4a Mandibular Artery
5 Maxillary Artery
5a Ophthalmic Artery
6 Temporal Artery*
6a Superficial Temporal Artery*
6b Frontal Branch of Temporal Artery
6c Occipital Artery
7 Basilar
8 Circle of Willis
8a Anterior Communicating Artery
8b Posterior Communicating Artery
9a Middle Cerebral Artery
9b Anterior Cerebral Artery
9c Posterior Cerebral Artery
9d Cerebellar

Veins

10 Superior Vena Cava
10a Subclavian Vein
11a Internal Jugular
11b External Jugular Vein
12 Vertebral Vein
13 Thyroid Vein
14a Superior Sagittal Sinus
14b Inferior Sagittal Sinus
14c Cavernous Sinus
14d Confluence Sinus
14e Transverse Sinus
14f Petrosal Sinus
14g Straight Sinus
15 Facial Vein
15a Ophthalmic Vein
15b Maxillary
16 Temporal Vein
16a Superficial Temporal Vein
16b Frontal Vein
16c Posterior Auricular
17 Occipital Vein

*Pulse Point Available

Cardiovascular 8b- Vessels of the Cranium and Cerebral Cortex

Arteries

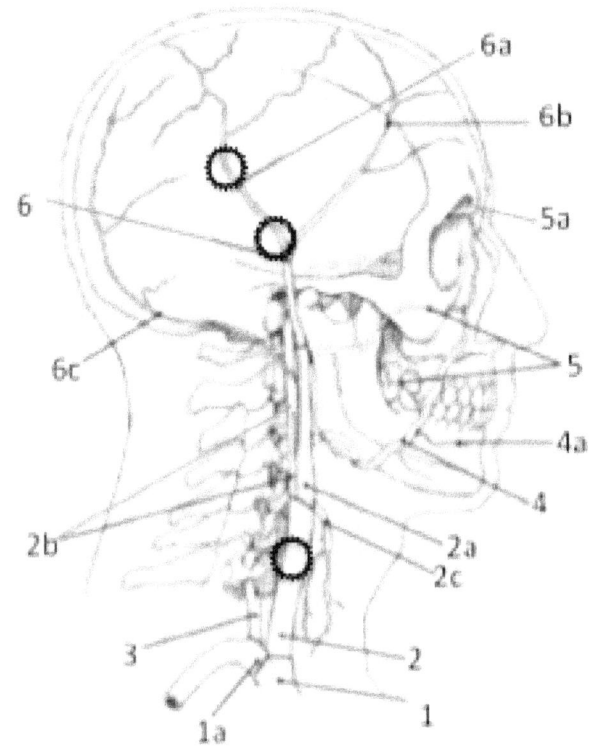

Veins

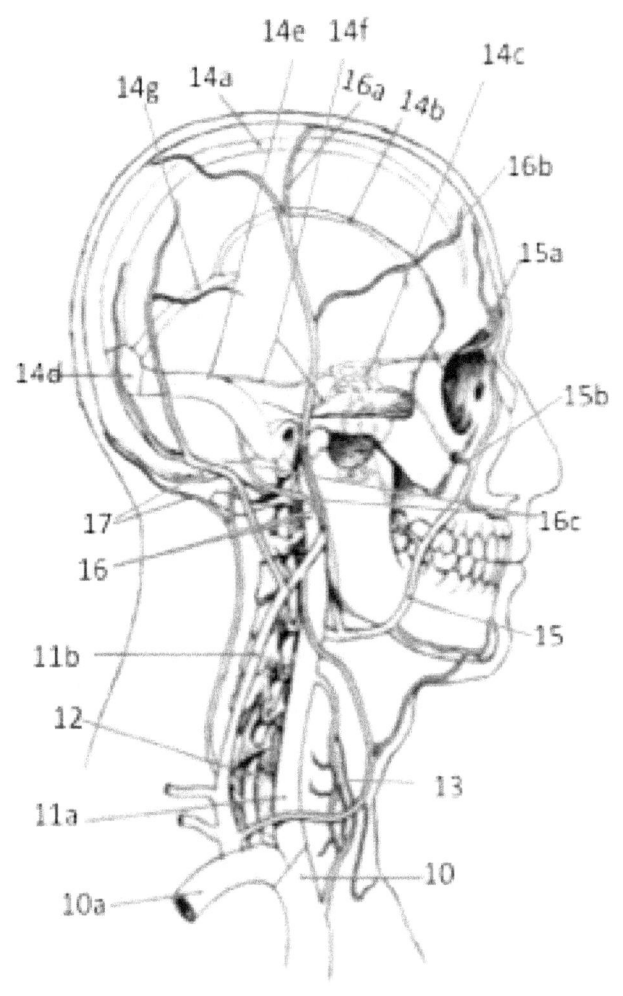

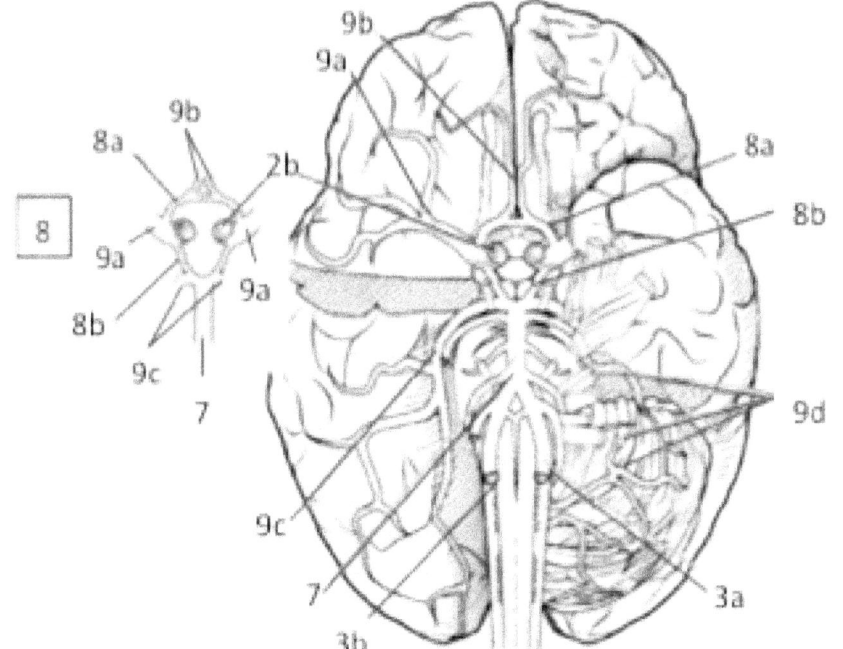

○ Pulse Point

Respiratory Overview

The respiratory system is comprised of two primary organs that are at the terminus of an open lumen (**ALVEOLI**) that provides an access for air movement from the external environment into the body. The pathway begins in the cranium (the nasal and oral cavity) and traverses through the ventral/anterior cervical region into the thoracic cage. The **UPPER TRACT** of the system is responsible for exchange of materials between the body and the external environment, defense against pathogens and harmful particles, and **VOCALIZATION** (speech and sound production) from specialized structures in the **LARYNX** (the region of the trachea between the Nasopharynx and the bronchiole branches). The **LOWER TRACT** of the system is responsible for passage of air from the environment to the alveoli, exchange of gasses between the alveoli and the pulmonary capillaries within the **LUNGS**. The pathway of air flow within the system is divided into the **TRACHEA, BRONCHI, BRONCHIAL BRANCHES, BRONCHIOLES,** and **ALVEOLI**.

While most people associate respiration (ventilation) with the need to "get Oxygen" for their body. **OXYGEN (O2) DOES NOT REGULATE** the system, **CARBON DIOXIDE (CO2)** or more importantly **BICARBONATE (H2CO3 or H+ HCO3-) REGULATES** the system. The action of the respiratory system is termed ventilation and is broken into two distinct components, the **INSPIRATORY (INHALATION)** and **EXPIRATORY (EXHILATION)** action that leads to the lungs **INFLATING** and **DEFLATING** (they do not expand and contract) proportionally to the movement the thoracic cage. This inflation and deflation is based on the changes in pressure within the thoracic cage leading to either a pressure less than the environment (inhalation) or pressure greater than the environment (exhalation). These changes in pressures are developed by the contraction and relaxation of the **DIAPHRGAM** that is impacted by **HEIGHT, AGE, LEVEL OF ACTIVITY, LEVEL OF ABDOMINAL FAT, HISTORY OF SMOKING** and **INTRA-ABDOMINAL VOLUME & PRESSURES**.

Respiratory Terms and Gas Laws

The physiological functions of the respiration (ventilation) is base on the combination of various chemistry and physical laws related to gasses in containers. Along with these laws, there are also concepts related to how gasses will move between and within regions of the body or between the body and the environment. Principal amongst this is:

DIFFUSION: the concept that individual components of a solution will move across a membrane based on the difference in concentration of that individual component between the two sides until the concentrations are equal

PARTIAL PRESSURE: the concept that each component of a solution will exert a percentage of the total pressure of the solution based on the percentage of that component within the whole solution

BOYLE'S LAW: Indicates that as volume of the thoracic cage increases the intrapleural pressures decrease allowing atmospheric gasses to diffuse to alveoli; as volume of the thoracic cage decreases the intrapleural increases allowing atmospheric gasses to diffuse to the external environment

FICK'S LAW: Details how the rate of diffusion of a substance is proportional to the difference in concentrations and total surface area of the membrane, but inversely proportional to the distance over which diffusion must occur across the membrane

DALTON'S LAW: Indicates that bulk flow of gasses within the cardiorespiratory system and exchange will based on the rate at which the gasses diffuse at the points of interaction (alveoli/capillaries, capillaries/tissues, atmosphere/alveoli) within the system

HENRY'S LAW: Indicates that the most likely gas to dissolve will be found in the plasma, while the least likely gas to dissolve will be found bound to a carrier molecule

CHARLES' LAW: Details how a change in plasmal temperature will affect the amount of $O_2(g)$ that will be bound to hemoglobin and the amount of $O_2(g)$ that will be dissolved in the plasma based on the ability to "cram" more gas into a warmer plasma

GRAHAM'S LAW: Details that the rate at which gas molecules diffuse is inversely proportional to the square root of its density in a solution and describes why gasses diffusion in direction that they diffusion based on the location within the body that were are examining.

Respiratory 1 - Overview

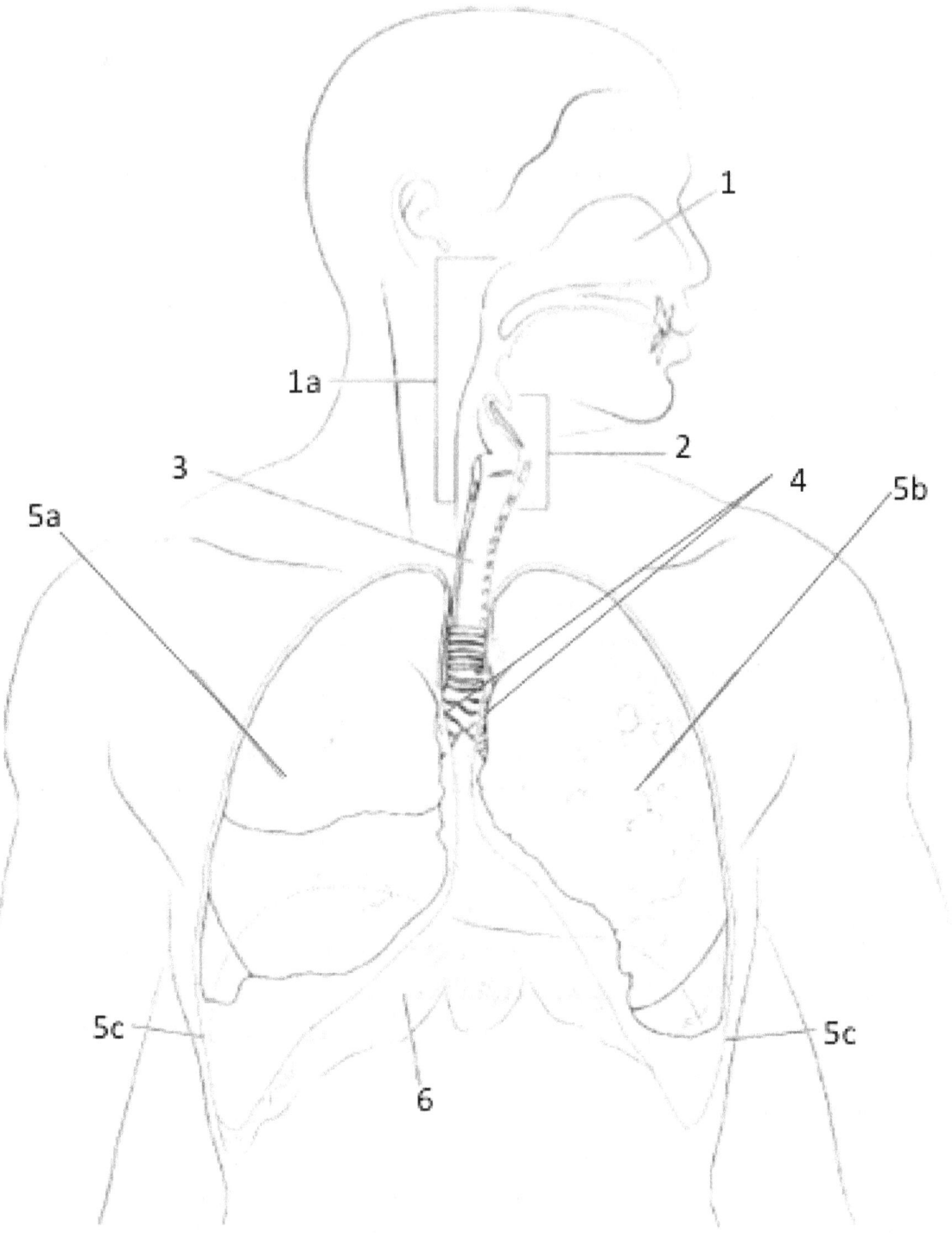

1 Nasal Opening
1a Nasopharynx
2 Larynx
3 Trachea
4 Bronchii

5a Right Lung
5b Left Lung
5c Pleural Cavity
6 Diaphragm

Respiratory 2a-Strucutres of the Systems

UPPER

1 Frontal & Nasal Sinus
2 Vomer
2a Superior Conchae
2b Middle Conchae
2c Inferior Conchae
3 Nares of Nose
3a Nasal Cavity
3b Cribiform Plate
4a Hard Palate
4b Soft Palate
4c Uvula
5 Epiglottis
5a Glottis
6 Trachea
7 Hyoid
8 Thyroid Cartilage
9 Cricoid Cartilage
9a Cricothyroid Ligament
9b Conus Elasticus
10 Vocal Folds
10a Vestibular Folds
11a Corniculate Cartilage
11b Arytenoid Cartilage

LOWER

12 Bronchi
13a Primary Bronchial Branches
13b Secondary Bronchial Branches
13c Tertiary Bronchial Branches
13d Bronchioles
14 Right Lung
14a Upper Lobe Right Lung
14b Middle Lobe Right Lung
14c Lower Lobe Right Rung
15 Left Lung
15a Upper Lobe Left Lung
15b Lower Lobe Left Lung
15c Lingula of Left Lung
16a Pleural Parietal Lining
16b Pleural Visceral Lining
16c Pleural Cavity Space

Respiratory 2b- Upper Respiratory Tract

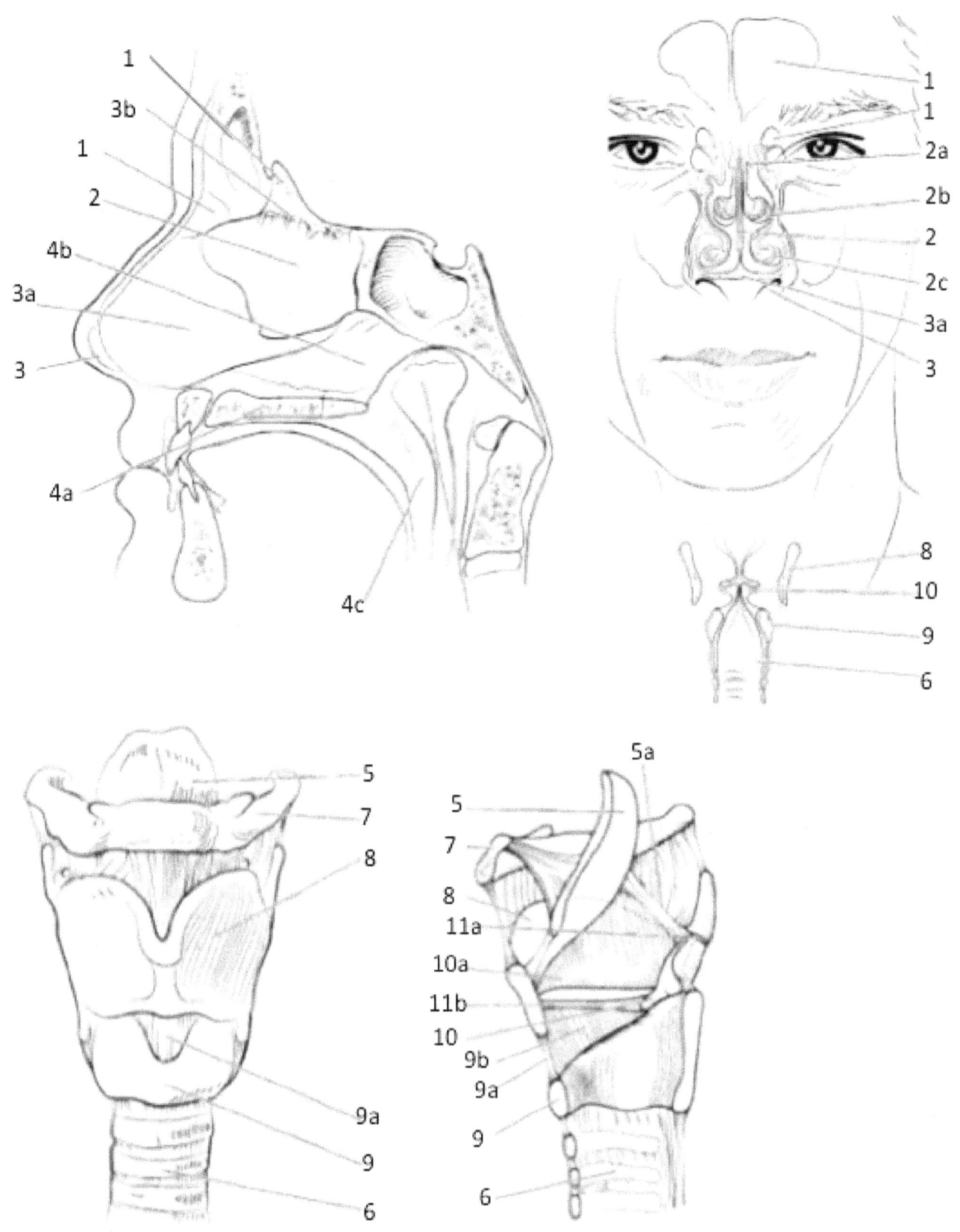

Respiratory 2C - Lower Respiratory Tract

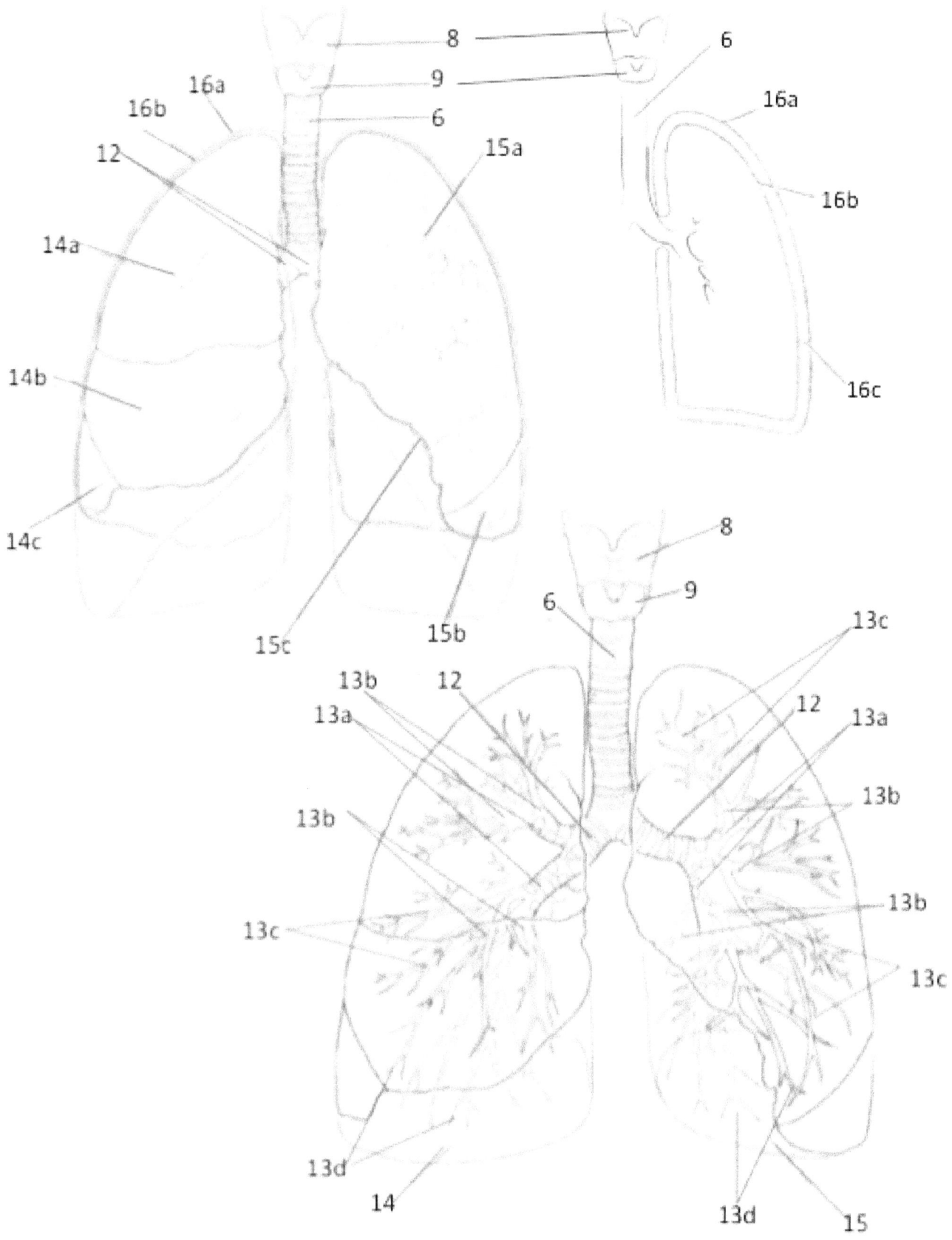

Respiratory 3- Alveoli

1a Bronchial Branches
1b Bronchioles
1c Alveolus
1d Alveoli
2 Primary Bronchii
2a Secondary Bronchii
3 Tertiary Bronchii
4a Alveolus Opening
4b Pulmonary Capillaries
4c Type I Alveoli Cells

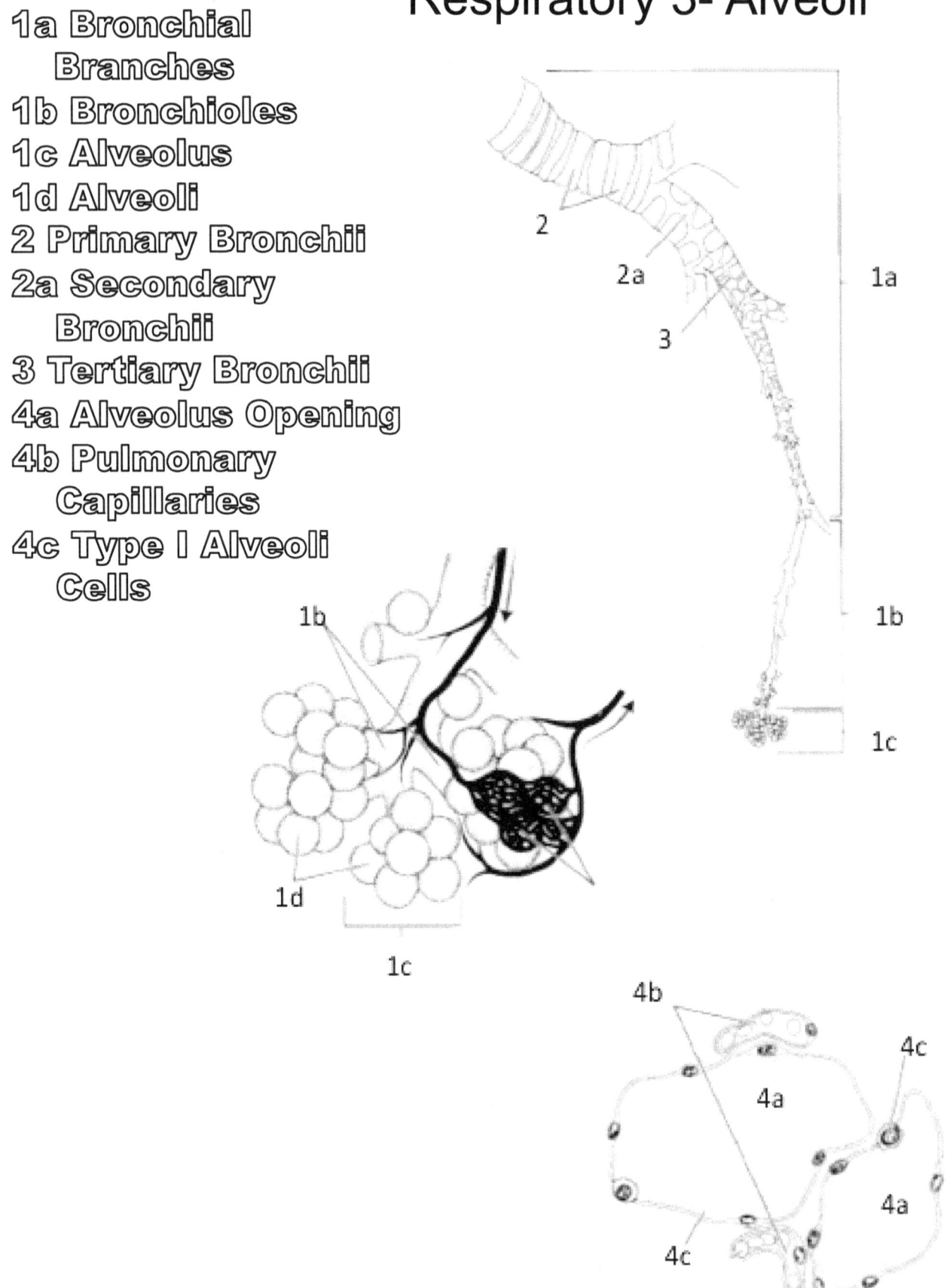

Respiratory 4a- Ventilation Cycle

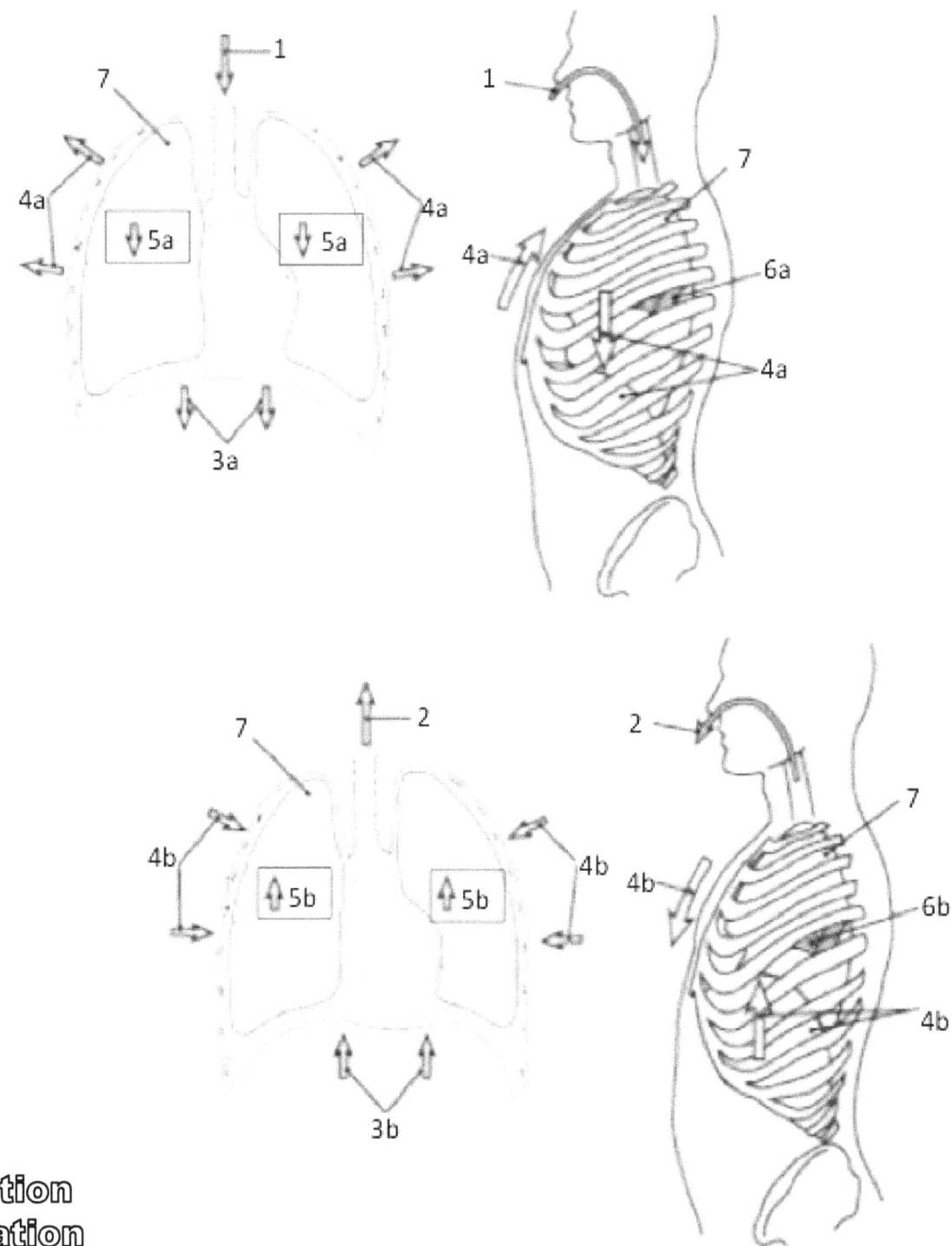

1 Inhalation
2 Exhalation
3a Diaphragm Contract
3b Diaphragm Relax
4a Expansion of Thorax Volume
4b Reduction of Thorax Volume
5a Decrease Pressure
5b Increase Pressure
6a Internal Intercostal
6b External Intercostal
7 Lung

Respiratory 4b- Ventilation Cycle Volumes

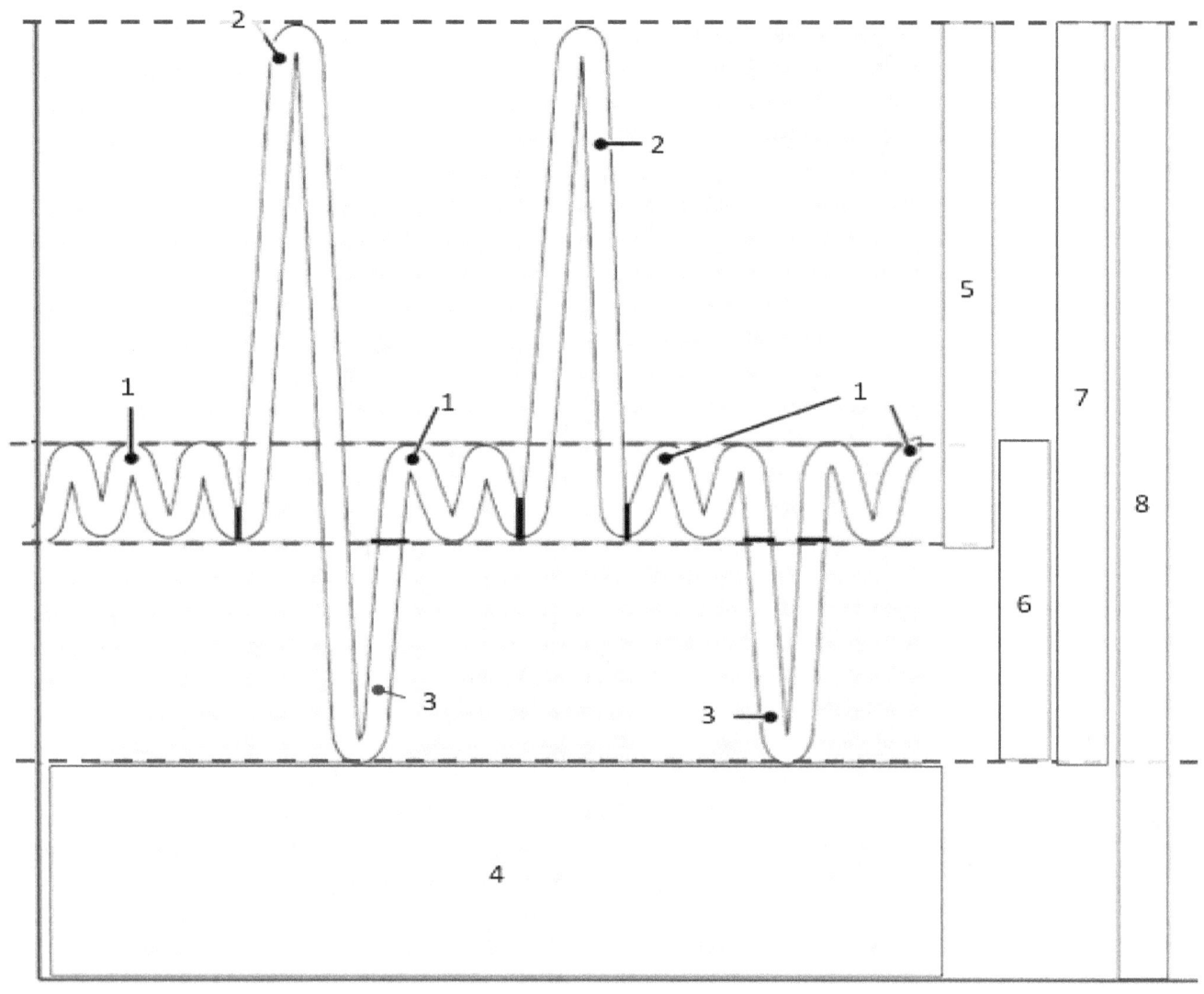

1 Tidal Volume, TV (~ 500 mL)
2 Inspiratory Reserve Volume, IRV (~ 3000 mL)
3 Expiratory Reserve Volume, ERV (~1000 mL)
4 Residual Volume, RV (~750-1000 mL)
5 Inspiratory Reserve Capacity, IRC=TV+IRV (~3500 mL
6 Expiratory Reserve Capacity, ERC=TV+ERV (~2000 mL)
7 Vital Capacity, VC=TV+IRV+ERV (~4500 mL)
8 Total Lung Capacity, TLC= VC+RV (~5000-6000 mL)

Gas Exchange- Oxygen (O_2) and Carbon Dioxide (CO_2)

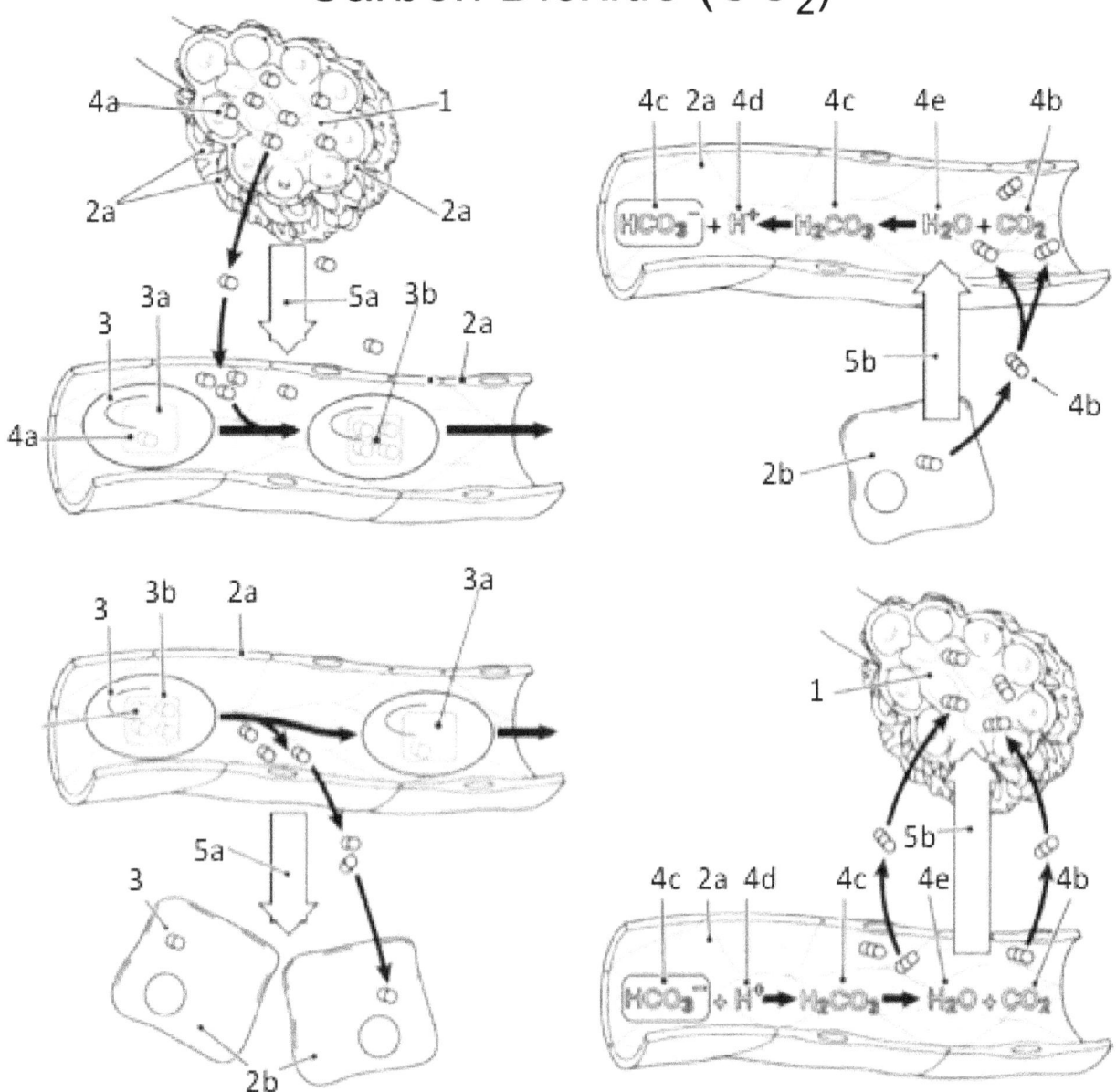

1 Alveolus
2a Capillaries
2b Tissues
3 Erythrocyte (RBC)
3a Hemoglobin
3b Oxyhemoglobin
4a Oxygen
4b Carbon Dioxide
4c Bicarbonate/ Bicarbonate Ion
4d Hydrogen Ion
4e Water
5a Diffusion Direction of O_2
5b Diffusion Direction of CO_2

Immune System

Barriers, Immune Cells, Vaccination, the protection against infection and response to injury

Overview of the System

The immune system is made of cells, tissues and organs that are set-up to protect an organism from **INFECTION and INFECTIOUS AGENTS** or **RESPOND TO INJURY OF THE BODY**. There are **SPECIFIC DEFENSES** and **NON-SPECIFIC DEFENSES** used by the system to keep infectious agents out of the body or fight the infectious agent once it enters the body. The methods of defense leads to the development of **IMMUNITY** for the individual. The concept of immunity is based on the ability to recognize what belongs to the body versus what does not belong. This recognition is developed by the presence or absence of distinct membrane proteins called **ANTIGENS**. If cell has an antigen that is not from the body it can be referred to as being a **PATHOGEN**, or an **INFECTIOUS AGENT**.

The non-specific defenses come from the **INNATE IMMUNITY** and include **STRUCTURES OF THE SKIN AND MUCUS MEMBRANES, SECRETIONS, LEUKOCYTES (IMMUNE CELLS) such as BASOPHILS, ESONOPHILS, NEUTROPHILS, MONOCYTES, MACROPHAGES and the NATURAL KILLER (NK) lymphocyte**, and **SWELLING**. The non-specific responses are the hall-mark responses that we typically label as having and **INFECTION** or "being sick" but are also what we see when ever we have injured ourselves or even the pain and burning that we feel from a workout. The most classic of the non-specific defenses is the **INFLAMMATION** response that is triggered from several hormones released following injury or infection, most important is **HISTAMINE** and **SEROTONIN**, and the puss that develops over an open wound. Many of these chemicals that cause inflammation are the underlying cause for many chronic diseases that people can suffer from.

The specific defenses come from the **ACQUIRED IMMUNITY** that develops within **B-LYMPHOCYTES (B-CELLS)** and **T-LYMPHOCYTES (T-CELLS)** that develops through the process of **IMMUNOCOMPETENCE**. Immunocompetence is the developed responses of T-cells and B-cells due to exposure to an infectious agent, or by being given a **VACCINE** to a known infectious agent, at various stages of the lifespan for the individual. The end result of exposure is that the person is either **IMMUNIZED** (previous exposed to pathogen) or VACCINATED (given the vaccine) for a distinct pathogen. Response to the immunocompetent system is based on the **IMMUNOGENICITY** (induce an immune response) or **IMMUNOREACTIVITY** (ability to have cell respond and react) of the T-cell, B-cell or other Immune cell to that pathogen.

The overall response is based on immunocompetence of the lymphocytes and immune cells to the pathogen that will allow for a **SPECIFIC RESPONSE** (elimination of the pathogen through a limited targeted immune and inflammatory response). This is different from the **GLOBAL RESPONSE** (elimination of the pathogen, or damage, through a systemic and wide ranging immune and inflammatory response) that is seen when the system is not competent to the pathogen, or following an injury. The global response is typically what is experienced when one thinks of a person being sick or having a disease. There are 2 types of responses within the immune/lymph system, both responses will involve the two lymphocytes, NK cells and the macrophages to "clean up the mess". Two ways being either the **ACUTE** injury, or infection (**< 72 hrs**) or **CHRONIC** injury, or infection (**> 72 hrs**), where the difference is in the novelty of the pathogen that is triggering the response, or the amount of tissue that has been injured. It is important to noted that the response cannot differentiate between **PSYCHOLOGICAL** and **PHYSIOLOGY INJURY/INFECTION**, because the response is a neuroendocrine response that is detailed in **SEYLE'S GAS RESPONSE** and inflammatory regulators, such as **INTERLEUKINS (IL'S)** and **CORTISOL**. Where if one is undergoing an inflammation response and can eliminate other stresses an immune response can be limited to being an acute, specific, response. Whereas, if the other stresses are increased can make an immune response that was acute and specific and turn it into a global and chronic response.

Immune 1-Lymph Vessels and Nodes of the Body

Vessels and Nodes

1 Thoracic Duct
1a Cisterna Chyli
2a Left Bronchomediastinal Trunk
2b Right Bronchomediastinal Trunk
3 Lymphatic Duct
4a Cervical Lymph Nodes
4b Parotid Lymph Nodes
4c Mandibular Lymph Nodes
4d Occipital Lymph Nodes
4e Jugular Trunks
5 Subclavian Trunks
6 Axillary Lymph Nodes
7 Mammary Lymph Vessels
8 Cubital Lymph Nodes
9 Mesenteric Lymph Nodes
10a Lumbar Lymph Nodes
10b Iliac Lymph Nodes
10c Inguinal Lymph Nodes
11 Femoral Vessels
11a Popliteal Lymph Nodes
12 Tibal Lymph Vessels
13 Superior Vena Cava
13a Subclavian Vein
13b Brachiocephalic Vein

Node Anatomy

14 Direction of Lymph Flow
14a Afferent Lymph Vessel
14b Valve
14c Efferent Lymph Vessel
14d Hilium
15a Germinal Centers
15b Medullary Cords
15c Trabeculae of the Node
15d Medullary Sinus
15e Cortical Nodules
15f Node Capsule

Immune 1a-Lymph Vessels and Nodes of the Body

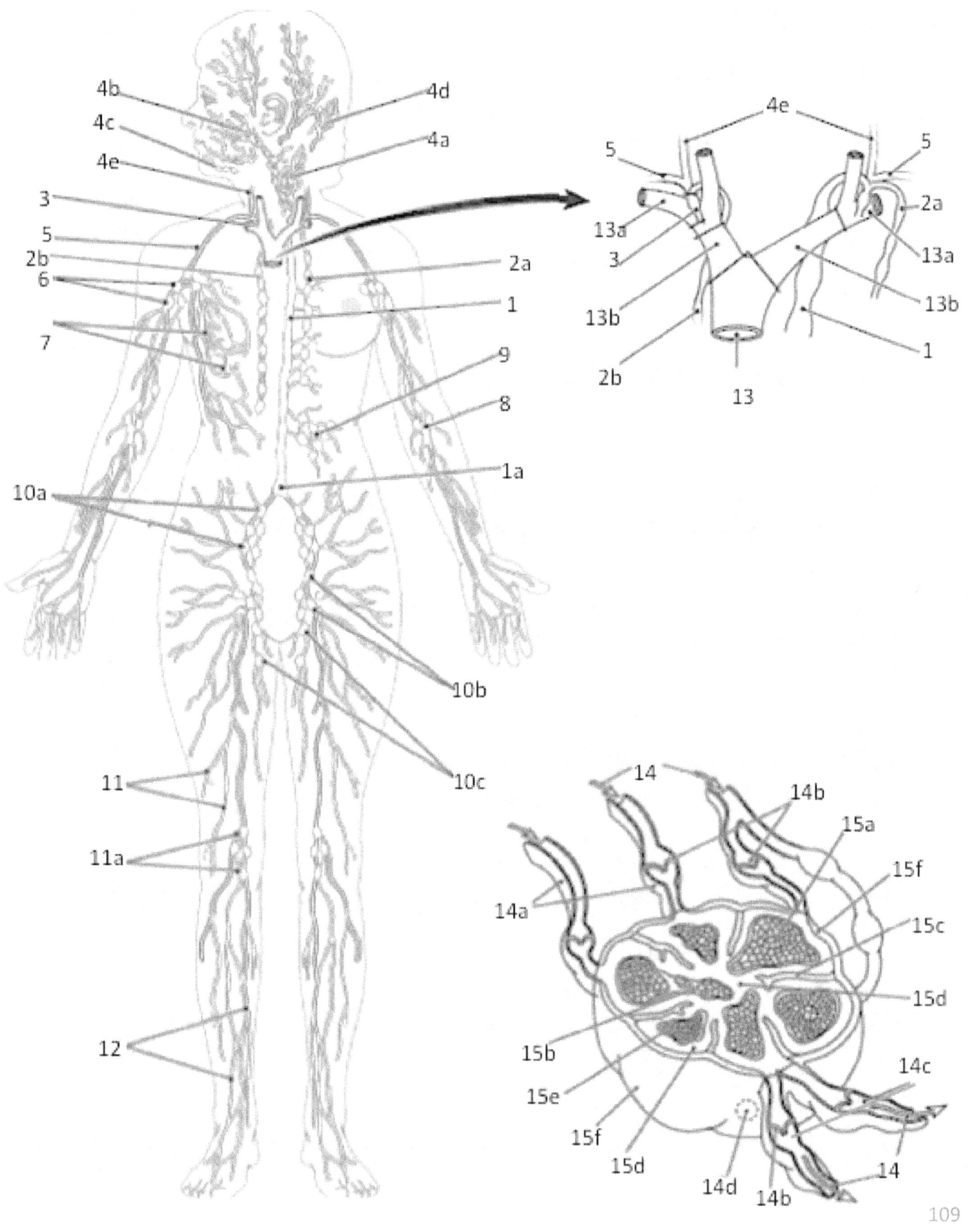

Immune 2-T-Lymphocyte (cell) Response

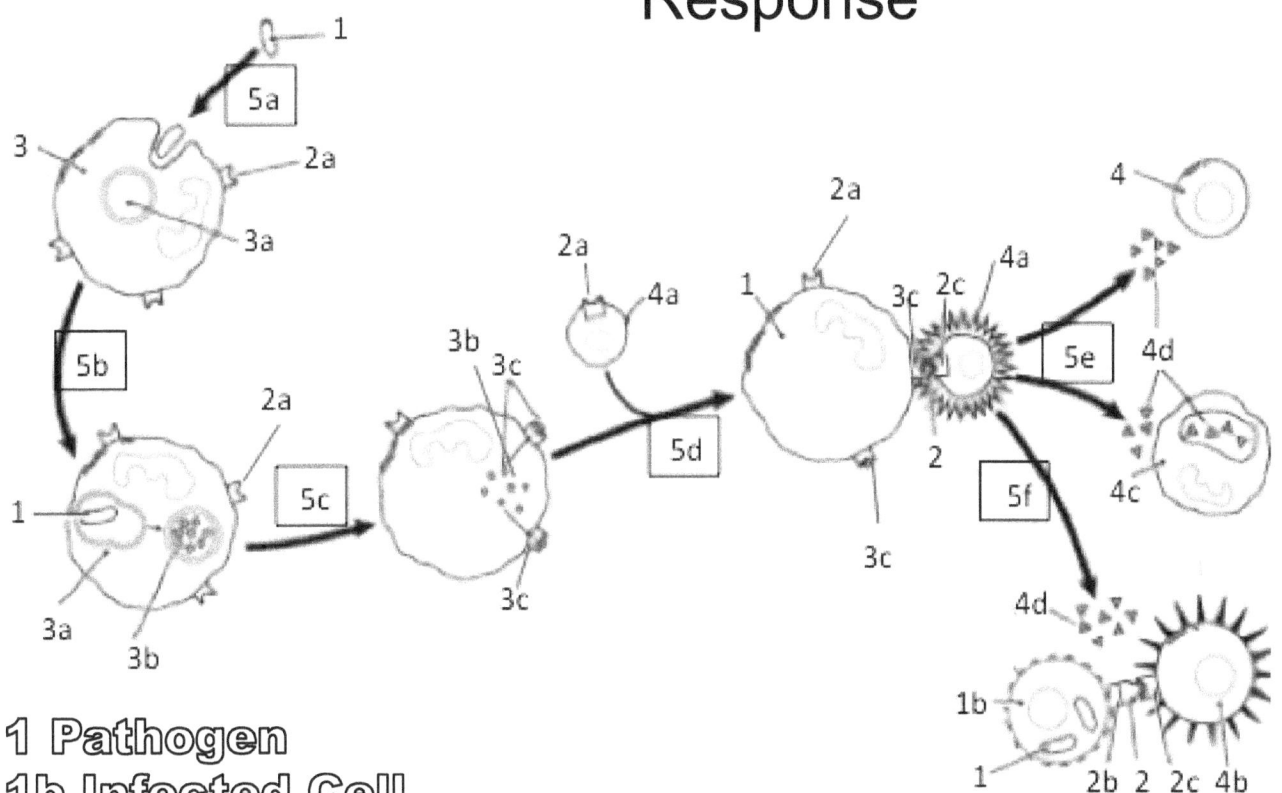

1 Pathogen
1b Infected Cell
2 Antigen Marker
2a MHC I
2b MHC II
2c CD-Factor Binding Site
3 Antigen Presenting Cell
3a Lysosome
3b Processed Antigen
3c Antigen Integration
4 T-cell
4a Helper T-Cell
4b Cytotoxic T-Cell
4c Antigen Presenting Macrophage
4d Interleukin
5a Pathogen Ingestion
5b Processing of Pathogen Antigen
5c Presentation of Antigen
5d Activation of Help T-Cell
5e Interleukin 2 (IL-2) Activation Lymphocyte & Macrophage
5f IL-2 Activation of Cytotoxic T-cell & Lysis of Pathogen

Immune 3a- B-Lymphocyte (B-Cell) Response

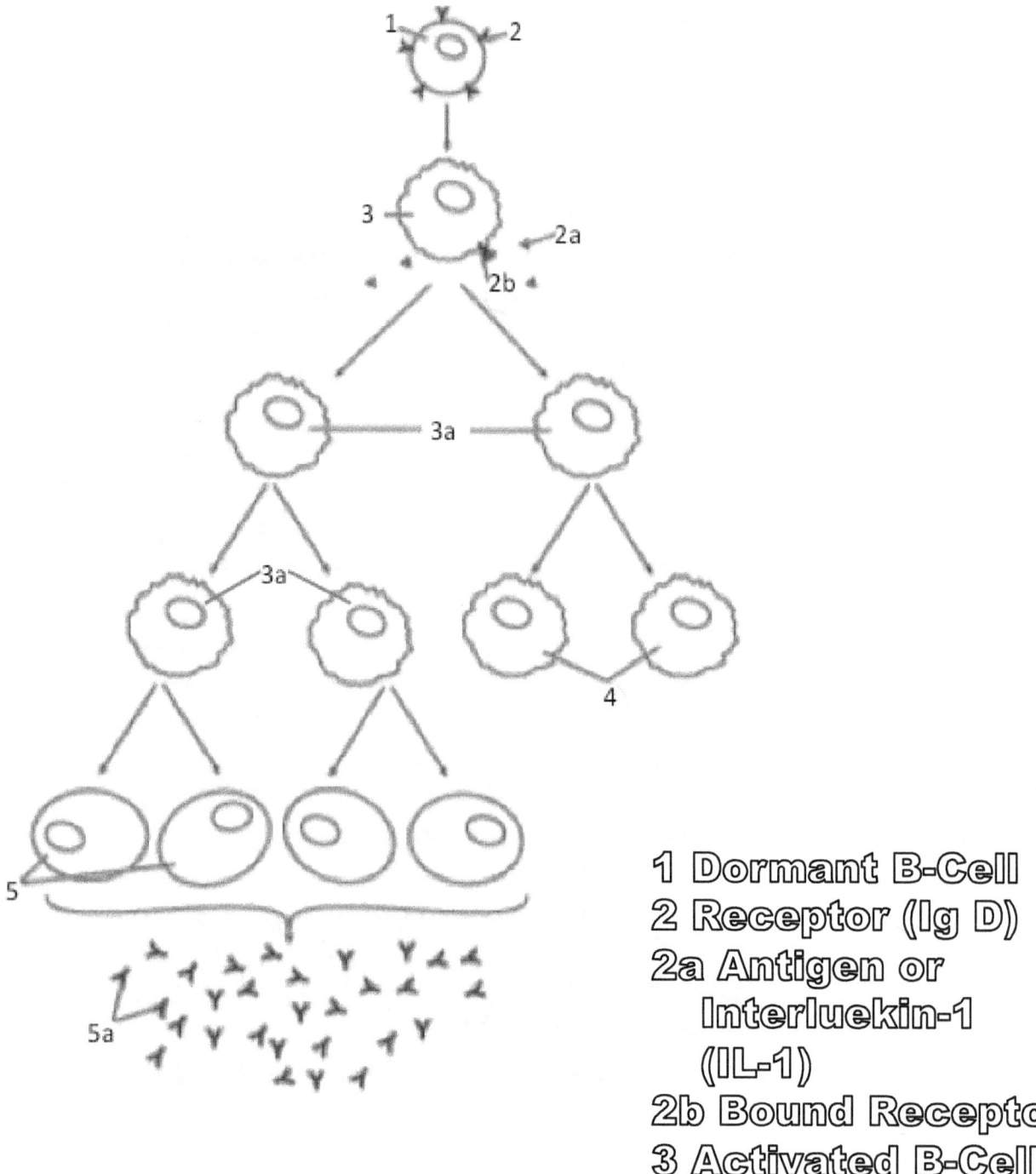

1 Dormant B-Cell
2 Receptor (Ig D)
2a Antigen or Interluekin-1 (IL-1)
2b Bound Receptor
3 Activated B-Cell
3a Cloned B-Cell
4 Sequestered (Memory) B-Cell
5 Plasmal B-Cell
5a Immunoglobin

Immune 4- Immunoglobulins (Ig)

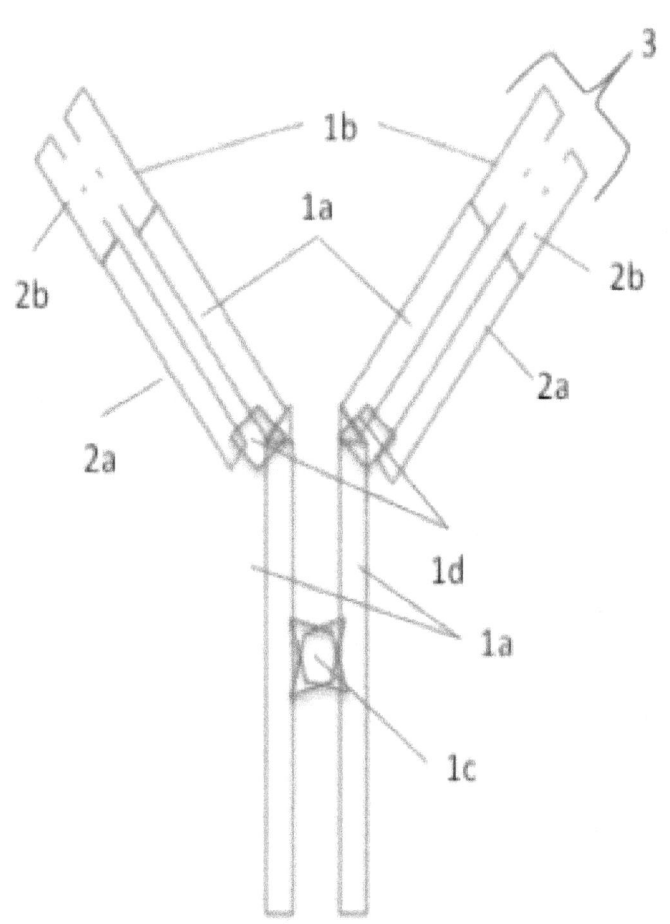

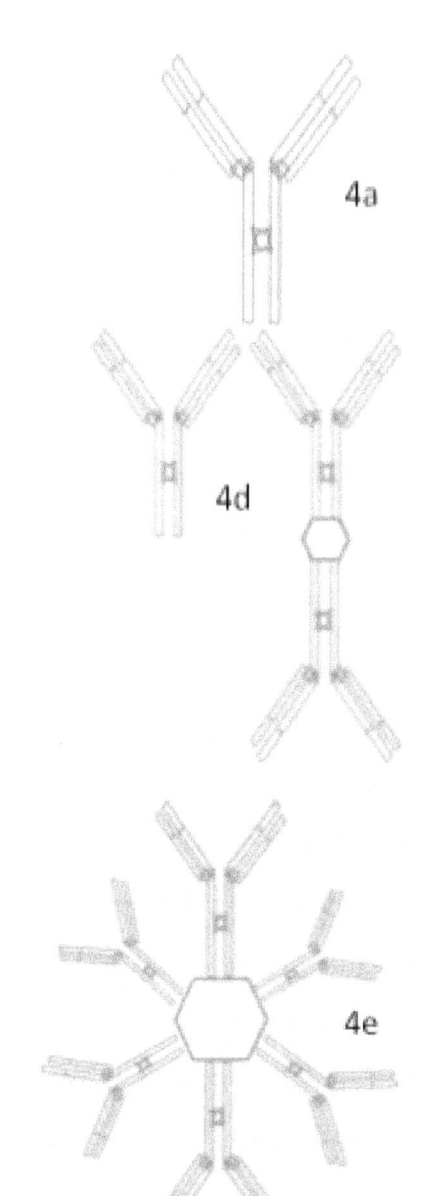

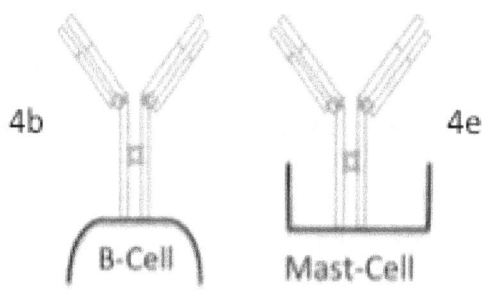

1a Conserved Heavy Chain
1b Variable Heavy Chain
1c Disulfide Bond
1d Hinge Point
2a Conserved Light Chain
2b Variable Light Chain
3 Antigen Binding Region
4a IgA
4b IgD
4c IgE
4d IgG
4e IgM

Immune 5- Allergen Response

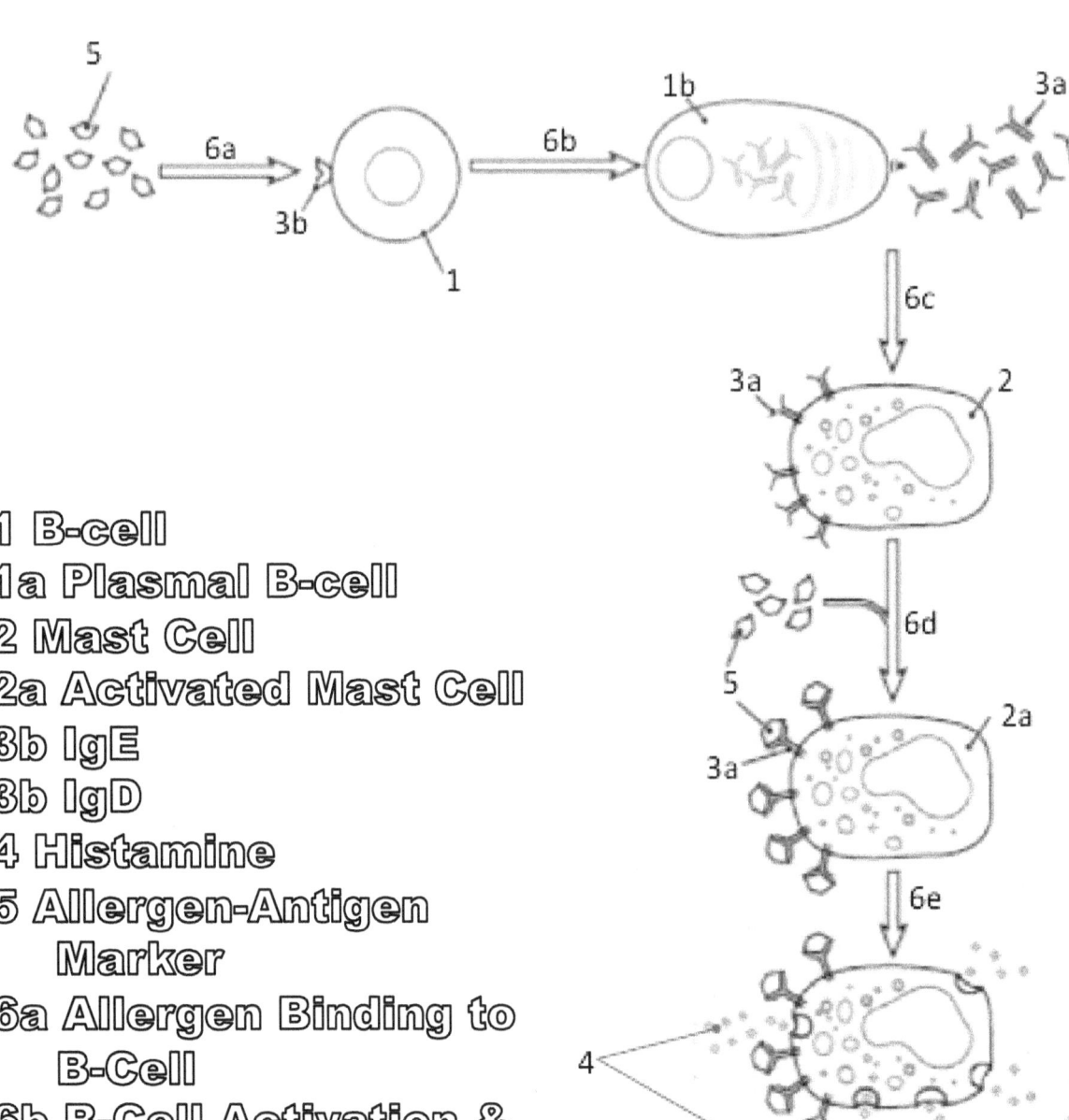

1 B-cell
1a Plasmal B-cell
2 Mast Cell
2a Activated Mast Cell
3b IgE
3b IgD
4 Histamine
5 Allergen-Antigen Marker
6a Allergen Binding to B-Cell
6b B-Cell Activation & IgE Release
6c IgE embedded in Mast Cell
6d Allergen Binding to Mast Cell
6e Histamine Release & Allergic Response

Vaccination Overview

Vaccinations are the principle means of developing a **PASSIVE ACQUIRED IMMUNITY**. Vaccination begins in early infancy after acquisition of immunoglobulins and other immune markers from one's mother in the form of **BREAST MILK**.

Following weaning, there are a series of vaccinations that the child will receive that acts as a new, or novel pathogen, but does not trigger any replication of the pathogen or extensive infection by the pathogen. This action of mimicking a pathogen triggers the acute immune response leading some to think that they are getting the disease. When in actuality, vaccines do not cause the disease to occur in the person getting vaccinated. All of the symptoms that person complains of simply immune system going through an initial global response so as to become "trained" by establishing memory cells to that pathogen so that immune cells can now identify the antigen and trigger memory cells. The end result of vaccination is the immunocompetence of the T-cells and B-cells for the individual.

The resulting immunocompetence limits likelihood of getting into a global response and instead are able to respond based on specific response. This ability to trigger specific immune responses limits pathogenic response & disease state from exposure to the pathogen that have been vaccinated against. **IT IS IMPORTANT TO REMEMBER THAT VACCINATION DOES NOT ELIMINATE THE PATHOGENIC ENTITY ONLY ELIMINATES THE ABILITY FOR THE INDIVIDUAL TO ACT AS A HOST FOR THAT PATHOGENIC ORGANISM**. This then reduces length of infection, limits "spread" to other individuals within close proximity to the infected individuals. As more members of a population get immunized and vaccinated to a specific pathogen, a **HERD IMMUNITY** develops within the population. This herd immunity ensures that if an infected individual should migrate into the population, the spread of a pathogen is limited within population as whole.

Immune 6-Herd Immunity

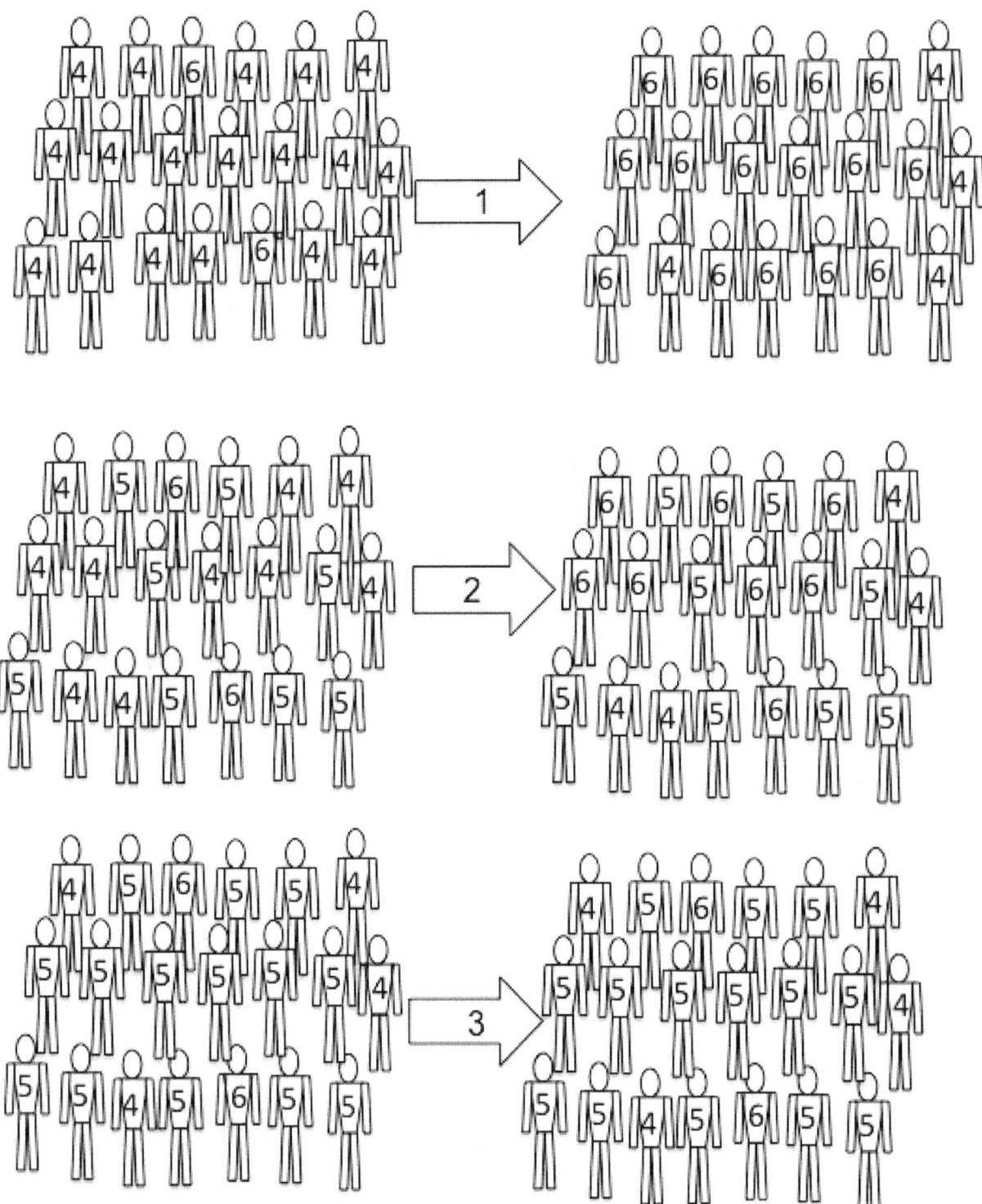

1 Infection without vaccination
2 Infection with partial vaccination
3 Infection with vaccination

4 Unvaccinated
5 Vaccinated
6 Infected

Gastrointestinal System

Overview of the System

The gastrointestinal system has been refined to transfer organic nutrients, minerals and water from the external environment into the internal environment. This process is specialized within the various organs and tissues within the system with a **FOUR-STAGE PROCESS** utilized in order to accomplish the general function based on the interaction between **THREE DISTINCT SEGMENTS** of organs. The entire process takes places within an elongated opening within the body that while contained with the body, encapsulates a lumen (**ALIMENTARY CANAL**) that is can be thought of as being external to the body. The **LENGTH (28-FEET) AND HISTOLOGY (HIGH LEVEL OF SURFACE AREA)** ensures complete digestion and absorption of all materials that enter the lumen within a approximately 24-hours following initial digestion in the mouth.

The **FIRST-STAGE** is the **MECHANCIAL DIGESTION**, taking the large bolus and breaking it into smaller bolus prior to entering the stomach, and takes place within the first segment of the system (**the ORAL OPENING**) **via MASCIATION** (use of **TEETH** and muscles of jaw). The end of mastication results in **SWALLOWING** and the initiation of **PERISTASIS**, the coordinated contraction of muscles (**CIRCULAR and LONGITUDINAL SMOOTH MUSCLES**) of the **ALIMENTARY CANAL** to propel food materials from the **MOUTH** to the **ANUS** in roughly a 24-hour period of time.

The **SECOND-STAGE** is the **TRANSPORTATION and INITIAL CHEMICAL DIGESTION** of the food material. This stage occurs in the second segment (**ESOPHAGUS and STOMACH**) where the smaller food boluses that were swallowed is then exposed to a low pH (ACIDIC) solution liquefies the material. This liquefaction is necessary for the **THIRD-STAGE** in the next segment (**SMALL INTESTINES**) for the further **CHECMIAL DIGESTION BY SPECIFIC ENZYMES** into the individual components (e.g. **FATS, CARBOHYDRATES, AMINO ACIDS, INORGANIC AND ORGANIC SALTS**). The **FOURTH-STAGE** takes place in the **SMALL and LARGE INSTESINES** and is **RESPONSIBLE FOR THE ABSORPTION OF THE INDIVIDUAL FOOD COMPONENTS** that have been digested so far within the system. This final segment is also responsible for the reabsorption of water and removal of undigested and nondigestible materials out of the system.

The anatomy of **THE ALIMENTARY CANAL (ESOPHAGUS, STOMACH, SMALL and LARGE INTESTINES)** and **ACCESSORY ORGANS (LIVER, GALL BLADDER and PANCREAS)** of the system function to maximize the amount that can be digested and absorbed based on blood flow through the tissues. Most organs will function based solely on the ingestion of materials and are not regulated to alter the amount of any individual type of material that can be absorbed. Any material that is not digested (**NON-DIGESTABLE**) or incompletely digested will not be absorbed and will be removed as fecal matter via **DEFECATION**, along with any excretions in the lumen. Of the material in the lumen, only 1% will be excreted in the form of fecal matter. The rate at which materials move through the system is based on the regulatory that the material is eaten, the amount that is eaten and the level of hydration. For most meals, materials that will be digested, absorbed and fecal waste removed within a 24-hour period of time.

The system is also involved with the regulation of when and how much we eat via the coordinated signals from hormones (**LEPTIN, GHRELIN, PANCREATIC HORMONES (PYY, INSULIN, GLUCAGON), NERUOHORMONES (NPY, POMC, ENDOCABINOIDS), GUT HORMONES (GASTRIN, SECRETIN, CCK), AND AMPKINASE**) along with stretch receptors within the alimentary canal that either initiates the **PULL-TO ("I NEED TO EAT")** or **PUSH-AWAY ("I'M FULL") FEEDING RESPONSES** through signals from the **HYPOTHALAMUS**.

Gastrointestinal 1- Overview

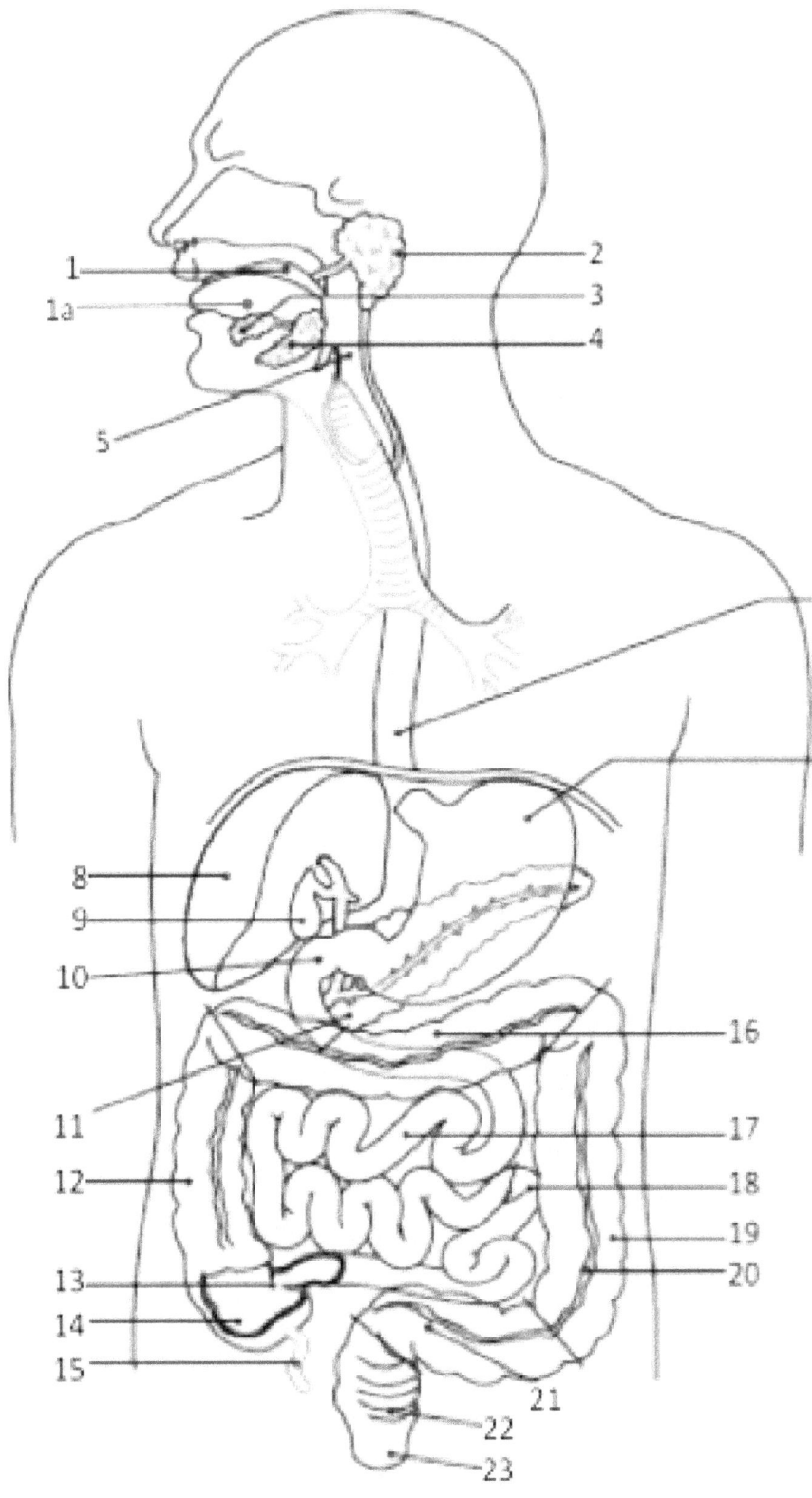

1 Oral Opening
1a Tongue
2 Parotid Gland
3 Sublingual Gland
4 Submandibular Gland
5 Glottis/Epiglottis
6 Esophagus
7 Stomach
8 Liver
9 Gall Bladder
10 Duodenum
11 Pancreas
12 Ascending Colon
13 Ileocecal Sphincter
14 Cecum
15 Appendix
16 Transverse Colon
17 Jejunum
18 Ileum
19 Descending Colon
20 Tenia Coli
21 Sigmoid Colon
22 Rectum
23 Anus

Gastrointestinal 2a- Oral Opening

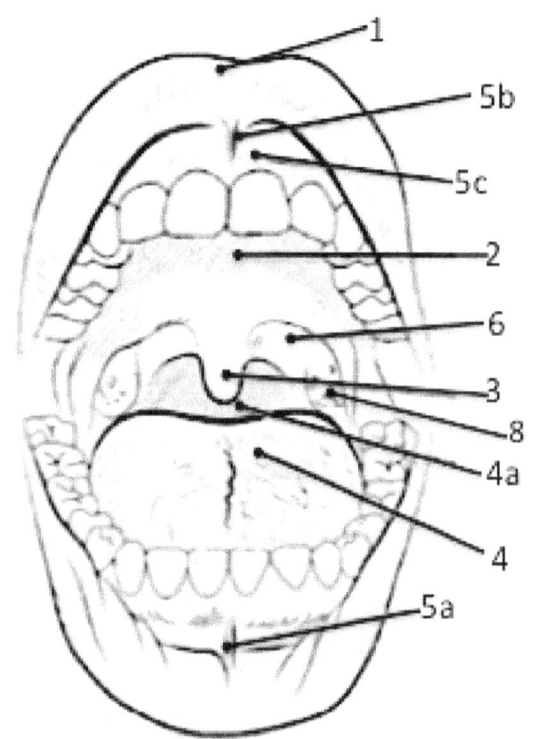

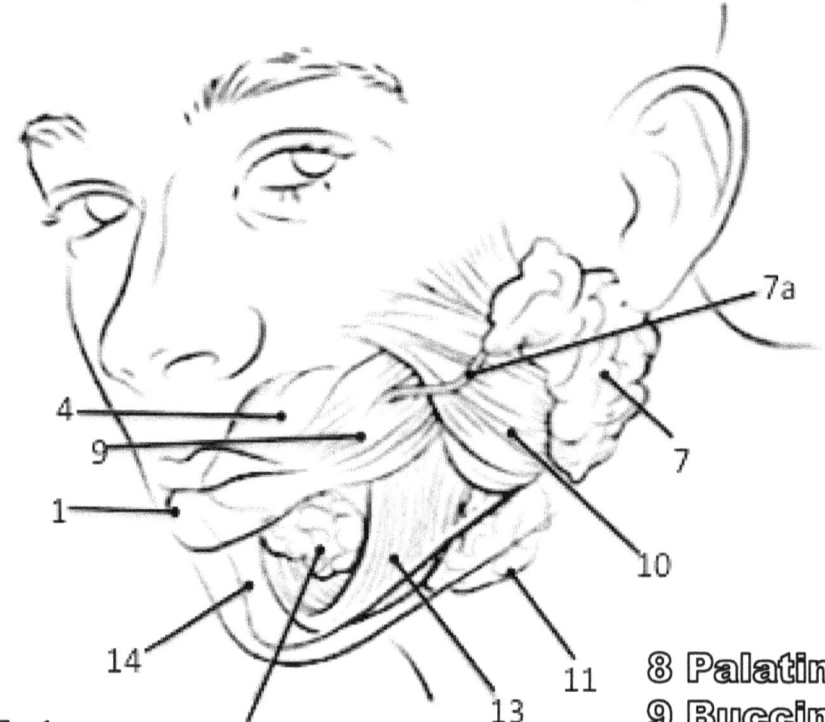

1 Lips
2 Hard Palate
3 Uvula
4 Tongue
4a Esophagus
5a Inferior Labial Frenulum
5b Superior Labial Frenulum
5c Gingiva
6 Palatopharyngeal Arch
7 Parotid Gland
7a Duct of Parotid Gland
8 Palatine Tonsils
9 Buccinator
10 Masseter
11 Submandibular Gland
12 Sublingual Gland
13 Mylohyoid
14 Mandible

Gastrointestinal 2b- Oral Opening: Teeth

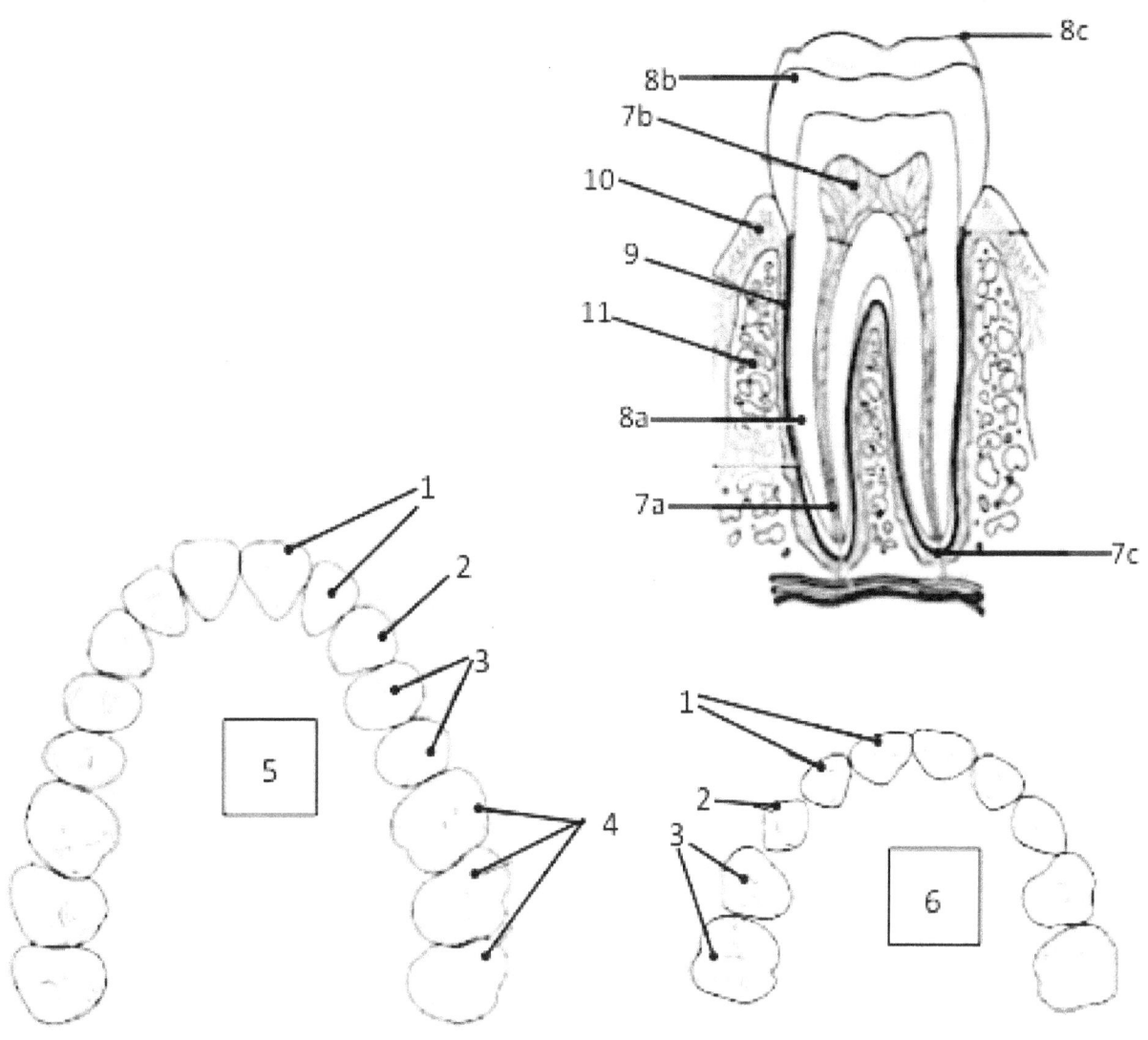

1 Incisor
2 Canine (Cuspid)
3 Pre-Molar (Bicuspid)
4 Molar
5 Permanent (Adult)
6 Deciduous (Juvenile)
7a Root Canal
7b Pulp Center
7c Blood Vessel & Nerve
8a Tooth Root
8b Tooth Crown
8c Tooth Enamel
9 Periodontal Ligament
10 Gingiva
11 Alveolus for Tooth

Gastrointestinal 3- Alimentary Canal

1 Serosa
2 Longitudinal Muscle
3 Nerves
4 Lymph
5 Blood Vessels
6 Submucosa
7 Circular Muscle
8 Mucosa
9a Circular Folds
9b Villi
10 Peyer's Patches (Lymph node)
11 Goblet Cells
12 Ciliated Columnar Epithelial
13 Brush boarder (Microvilli)

14 Lactose
14a Lactase
15 Disaccharide
15a Disaccharidase
16 Polysaccharides
16a Amylase

17=Monosaccharides
18=Protein (Polypeptide)
18a=Peptidase
19=Amino Acids
20=Dipeptides
21=Lipid Droplet
21a=Bile
22=Micelle & Triglycerides
22a=Lipase
23=Moderate-Chain Fatty-acid
24=Short-Chain Fatty-acids
25=Chylomicron & Long-chain Fatty-aids

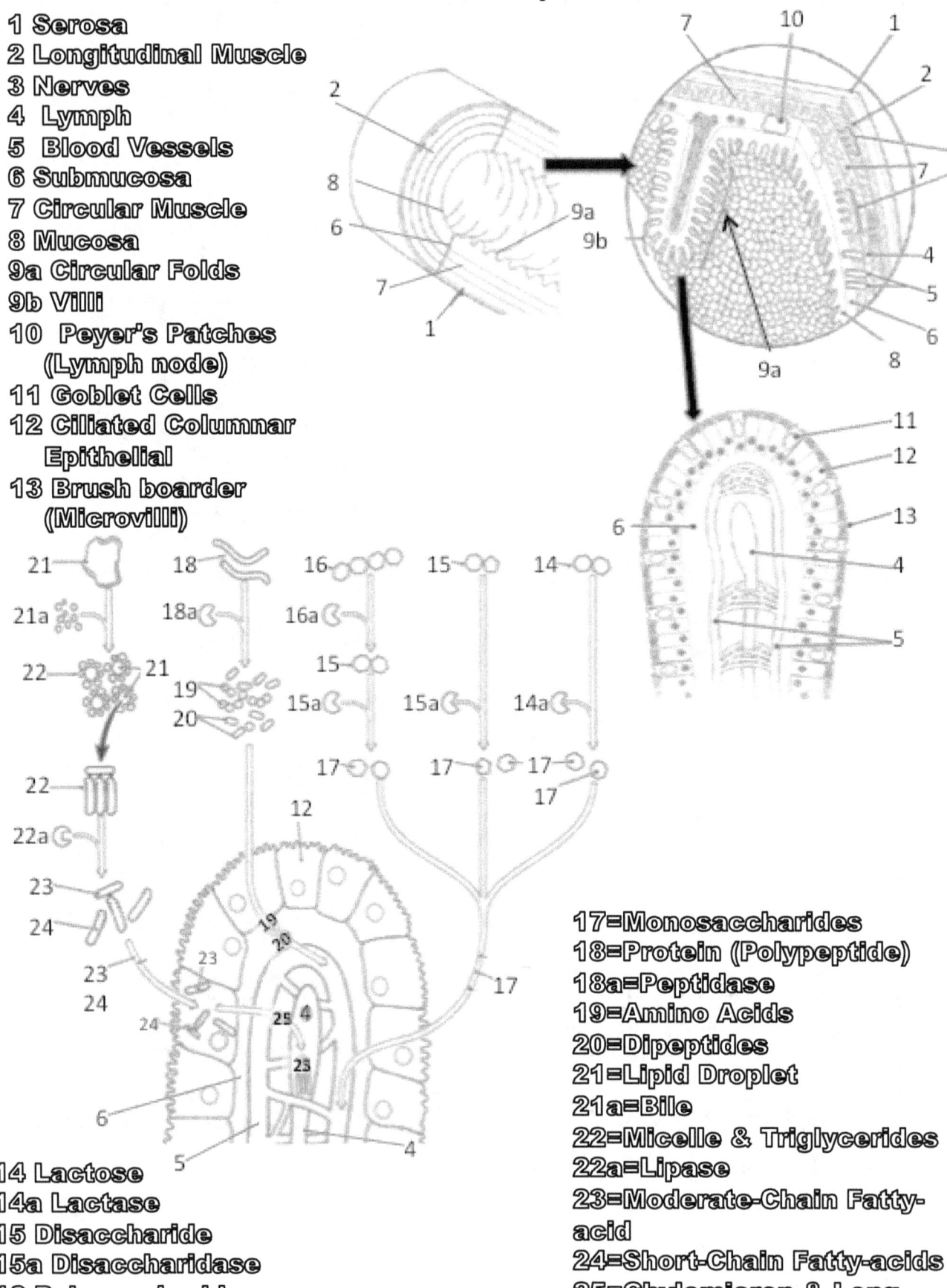

Gastrointestinal 4- Stomach

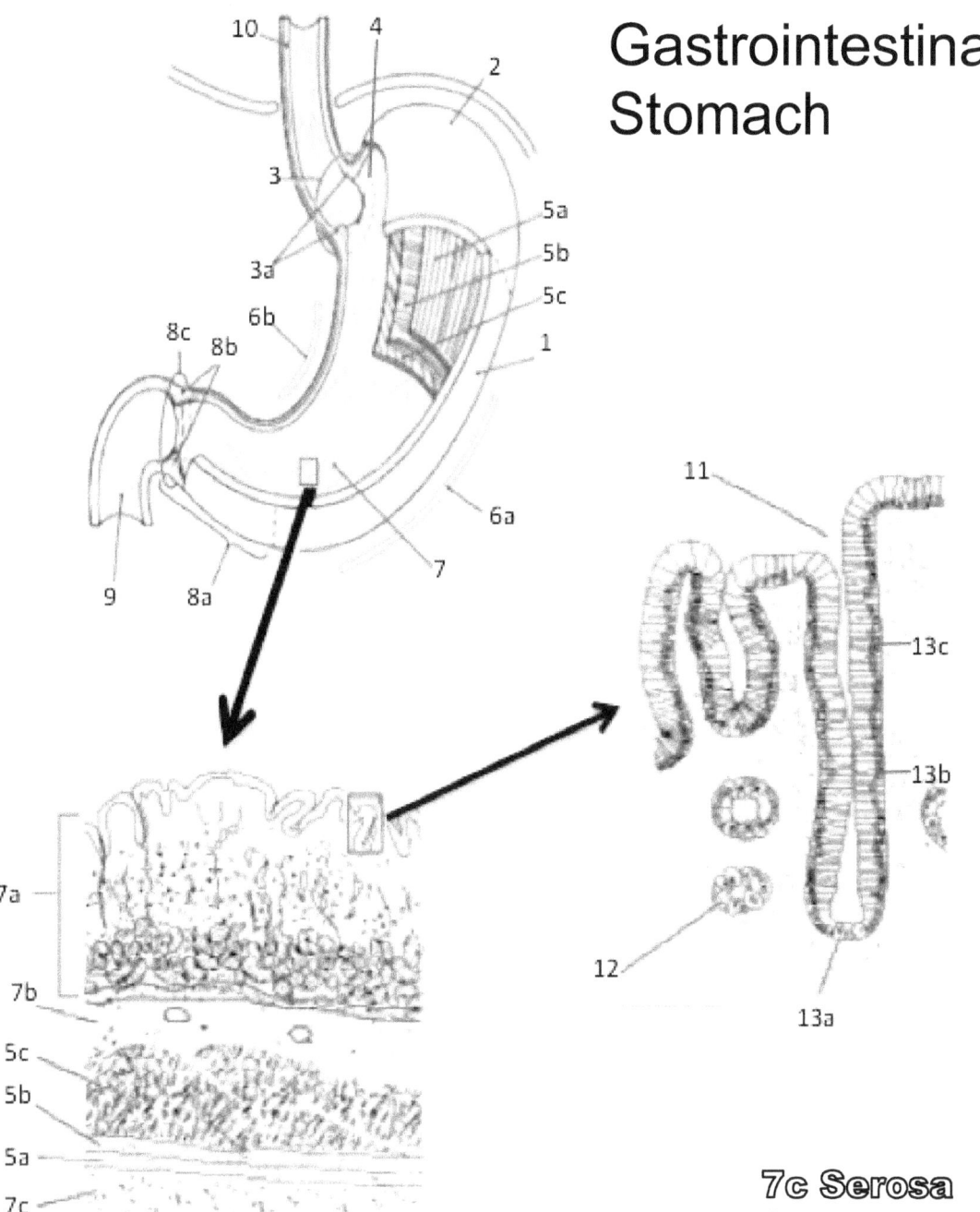

1. Body of the Stomach
2. Fundic Region
3. Upper Esophageal Sphincter
3a. Esophageal Sphincter
4. Cardiac Region
5a. Longitudinal Muscle
5b. Circular Muscle
5c. Oblique Muscle
6a. Greater Curvature
6b. Lesser Curvature
7. Rugae of Stomach
7a. Mucosa
7b. Submucosa
7c. Serosa
8a. Pyloric Region
8b. Pyloric Canal
8c. Pyloric Sphincter
9. Duodenum
10. Esophagus
11. Gastric Pit
12. Gastric Glands
13a. Enteroendocrine (G) cells
13b. Chief Cells
13c. Neck Cells

Gastrointestinal 5 - Small Intestines

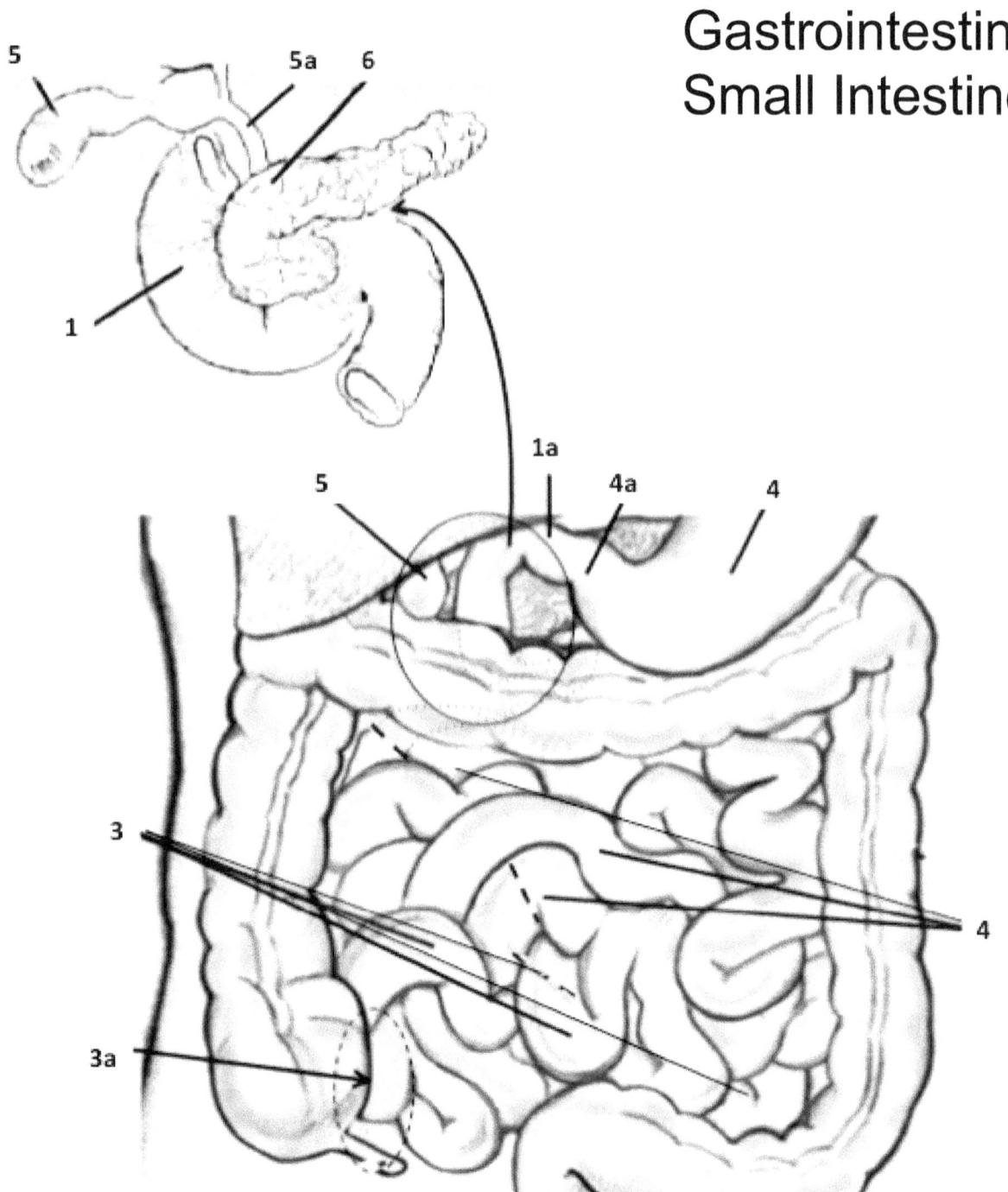

1 Duodenum
1a Pyloric Sphincter
2 Jejunum
3 Ileum
3a Ileocecal Sphincter
4 Stomach
4a Pyloric Canal
5 Gall Bladder
5a Cystic Duct
6 Pancreas

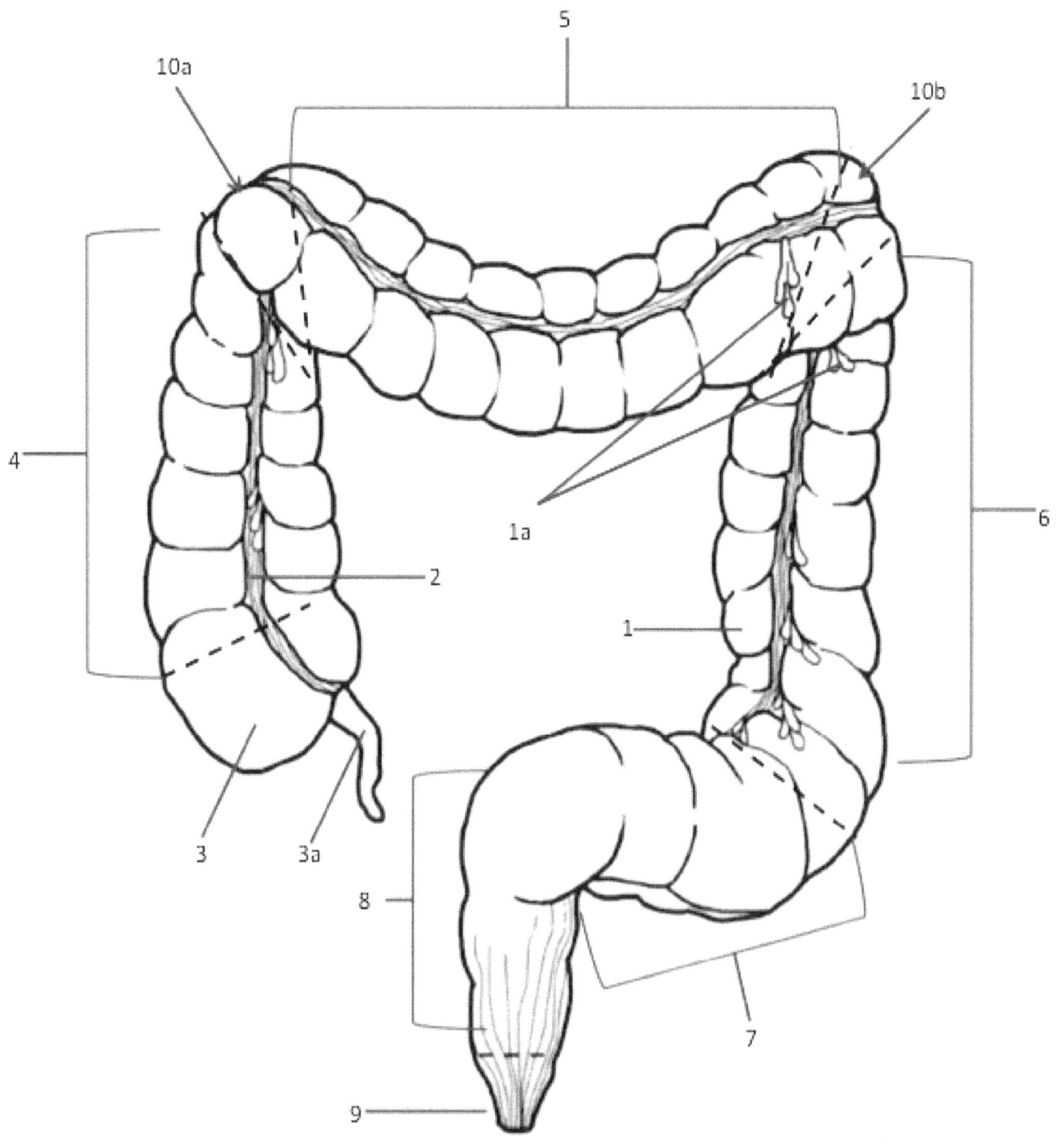

Gastrointestinal 6a - Liver

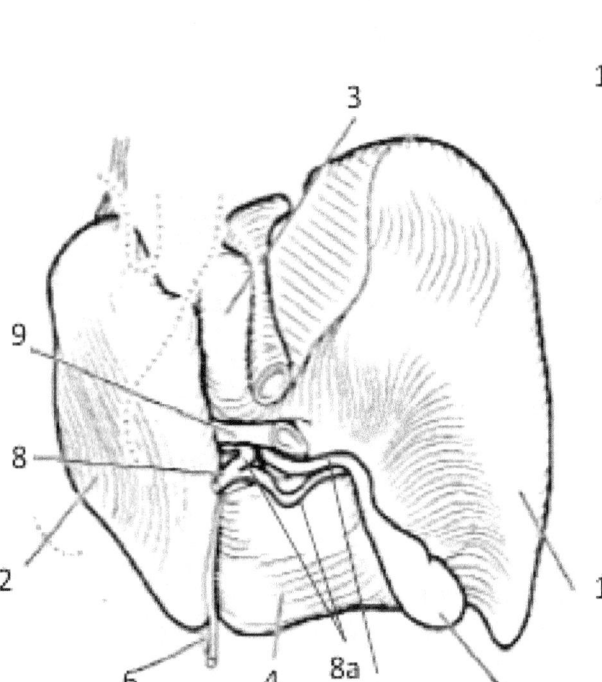

1 Left Lobe of Liver
2 Right Lobe of Liver
3 Quadrate Lobe of Liver
4 Caudate Lobe of Liver
5 Falciform Ligament
6 Round Ligament
7 Gall Bladder
8 Common Bile Duct
8a Hepatic Bile Ducts
8b Cystic Duct
9 Hepatic Portal Vein
10 Hepatic Triad
11a Branch of Hepatic Portal
11b Central Lobule Vein
12 Branch of Hepatic Artery
13a Hepatocytes
13b Sinusoid
13c Bile Canaliculi

Gastrointestinal 7- Pancreas and Bile Ducts

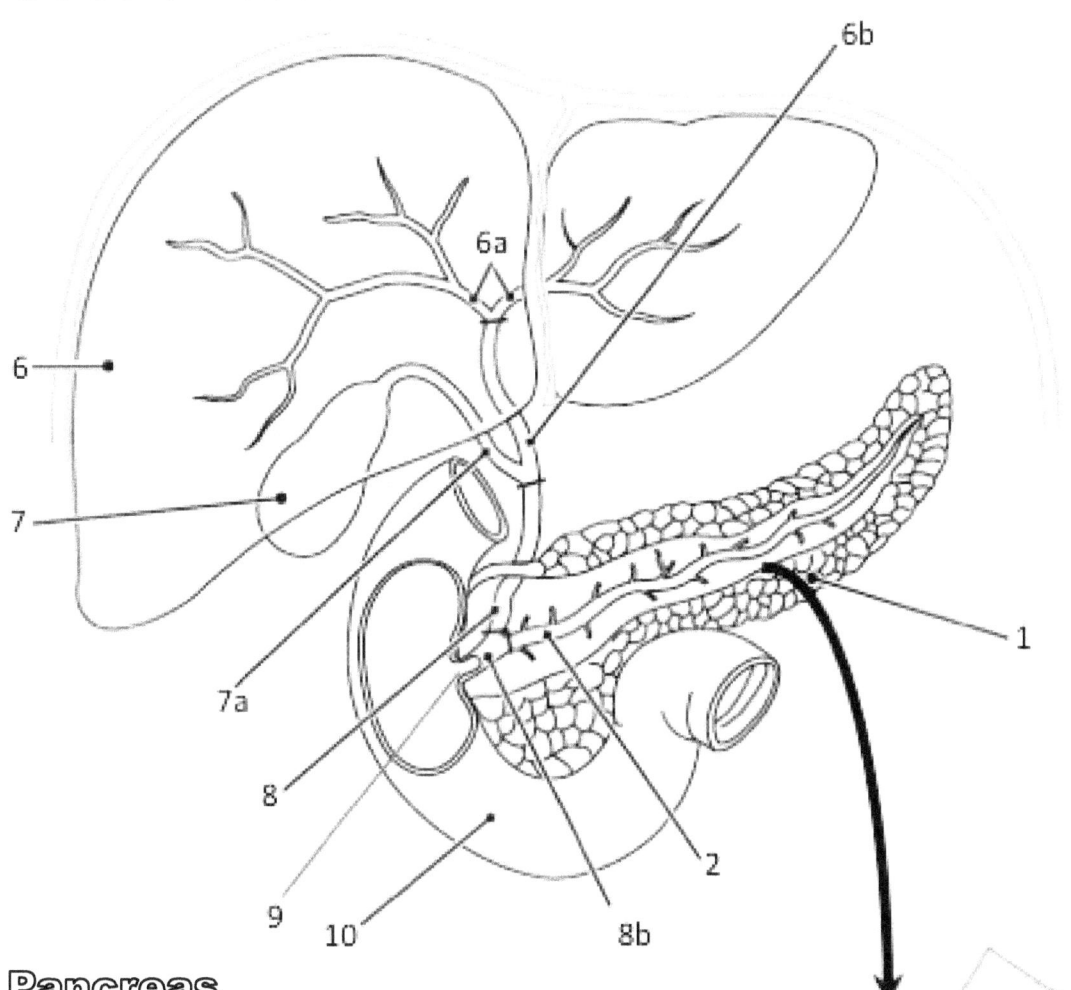

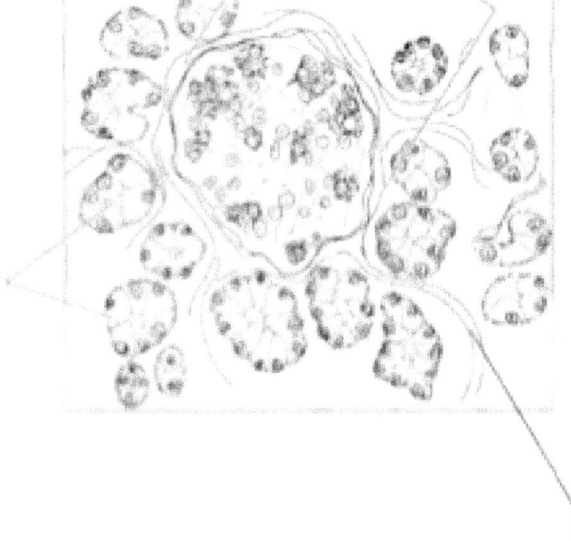

1 Pancreas
2 Pancreatic Duct
3 Acinar Cells
4 Islets of Langerhans
5 Capillaries
6 Liver
6a Lobular Bile Ducts
6b Hepatic Bile Ducts
7 Gall Bladder
7a Cystic Duct
8 Common Bile Duct
8b Ampulla of Hepatopancreatic Duct
9 Sphincter of Oddi
10 Duodenum

Excretory System

Kidney and Renal Functions & Water Balance Issues

Overview of the System

The excretory system is a complex system that integrates a number of organs (KIDNEY, LIVER, LUNGS, SKIN/SUDORIFEROUS GLANDS) from various regions of the body that all function together to TRANSFORM AND REMOVE TOXINS AND METABOLIC WASTES, REGULATE pH OF THE BODY, AND REGULATE HYDRATION and ELECTROLYTE BALANCE. The two principal sites of action are in the LOBULES of the LIVER (via HEPATOCYTES) and the NEPHRON of the KIDNEY.

System is responsible for a number of homeostatic regulation processes that ensure that plasma chemistry remains stable. That is done through MONITORING CONCENTRATIONS OF METABOLITES (substances produced in a reaction that are released from the site of origin, but can be used in metabolic reactions), EXCRETING WASTE PRODUCTS (substances produced in a reaction that are released from the site of origin that can no longer be used in metabolic reactions), TRNSFORMING AND EXCRETING ENDOTOXINS and EXOTOXINS (substances that are either produced within the body or ingested that can disrupt normal physiological functions), BUFFERING CHANGES IN pH (via accumulation or excretion of acids and bases that act as buffers to each other), and MONITORING LEVELS of ELECTROLYTES (ions responsible for membrane functions) and OSMOLARITY (concentration of solutes within the plasmal water).

Within the overall functions, the **liver** is responsible for the **removal of SOLID and AQUEOUS TOXINS, LIPIDS, and ANY SOLID WASTE PRODUCTSS (non-water soluble) from the blood**. The **lungs** are responsible for the **removal of GASSEOUS (VOLITILE) SUBSTANCES from the blood**. The **kidneys and sudoriferous glands** are responsible for the **removal of any AQUEOUS (WATER SOLUBLE) TOXINS and WASTE PRODUCTS from the blood**.

The system functions so that each organ in the system works in conjunction with, and independent of, the other organs. This interaction and independence ensures that blood chemistry remains relatively stable and within homeostatic ranges for the varies substances. Resulting in a plasmal **pH RANGE of 7.4-7.5 (average 7.42)** through the use of the buffer systems (**BICARBONATE, AMONIA, PHOSPHATES AND PROTEINS**) that can be used either as weak acids, or weak bases to quickly and temporarily bind H+ (do not remove H+). Typically the buffers will act as Lewis Bases (willing to accept H+) to the Lewis Acids (willing to give H+) in the aqueous environment of the cell or plasma until the H+ ions can be excreted from the body. Maintain both **WATER AND ION CONCENTRATIONS** within the tissues and fluids of the body at a homeostatic level so that membrane potentials are homeostatically maintained to keep functions of all cells normal and along with that hydration balance. That **COORDINATES HORMONE AND NERVES** signals from the the **HYPOTHALAMUS TO INITIATE THE DESIRE TO CONSUME WATER (DRINK)** or reabsorb water during the process of forming urine from the filtrate. Water and ion balance is also important in the buffering and pH balance of the body through coordinating the **ANION GAP**, difference between – IONS (including HCO_3^-, Cl^-) and + IONS (including Na^+, K^+, Ca^{++}) concentrations. These ions and water balance are are regulated hormonally by **ALDOSTERONE, VASOPRESSIN/ANTIDIURETIC HORMONE (ADH), ATRIAL NATRIURETIC PEPTIDE (ANP), ANGIOTENSIN II, CALCITONIN, CALCITROL, AND PARATHYROID HORMONE (PTH).**

Excretory 1-Overview

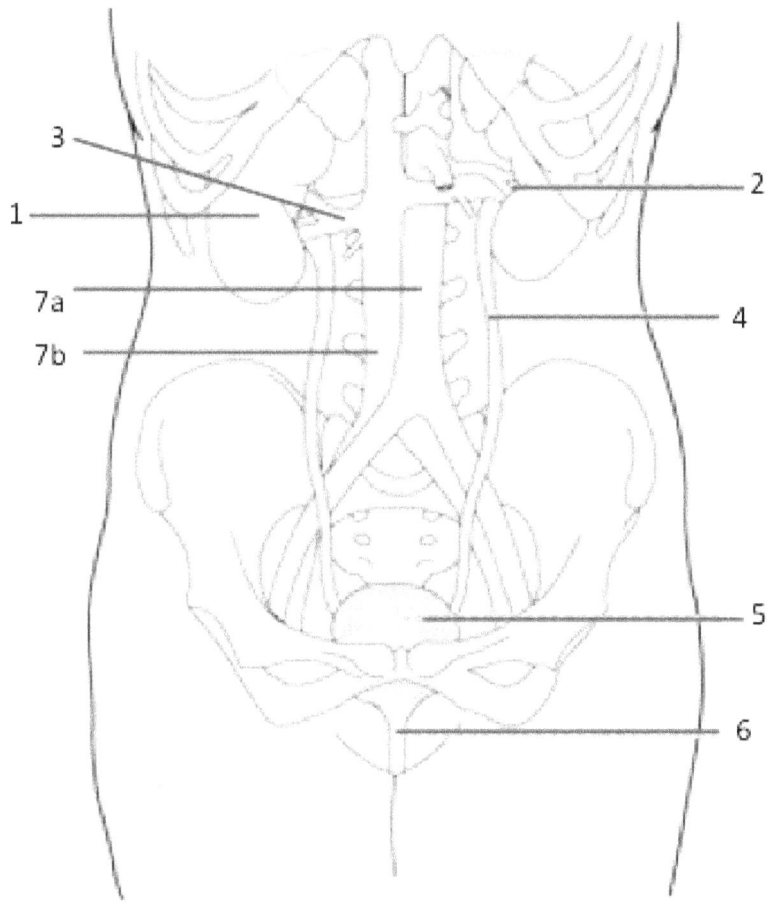

1 Kidney
2 Renal Artery
3 Renal Vein
4 Ureter
5 Urinary Bladder
6 Urethra
7a Aorta
7b Inferior Vena Cava

Excretory 2- Renal Anatomy

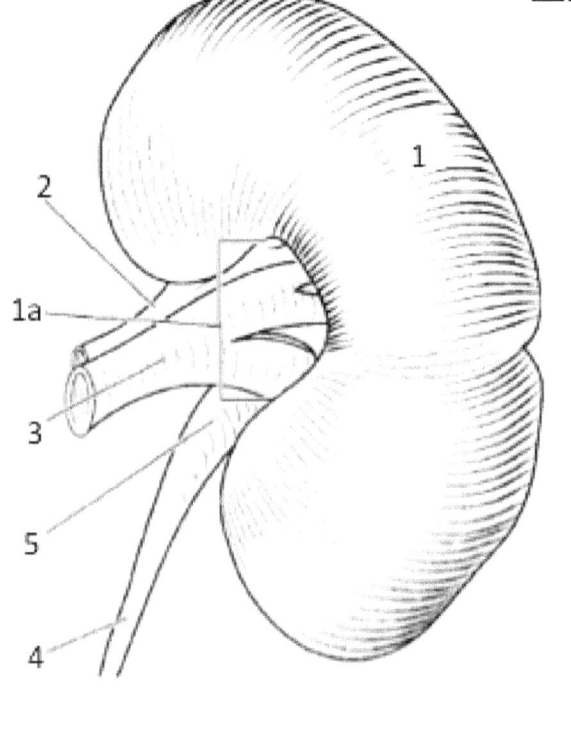
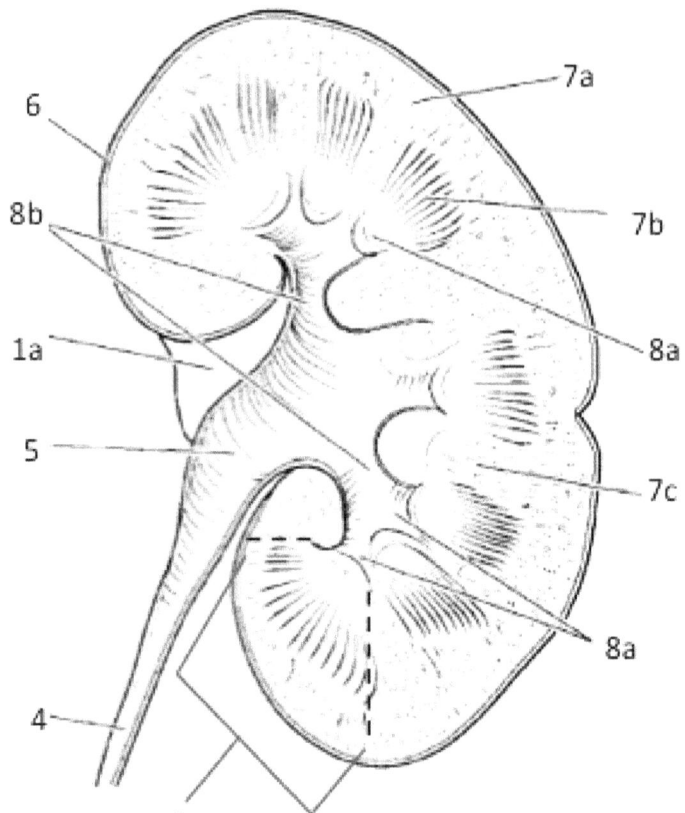
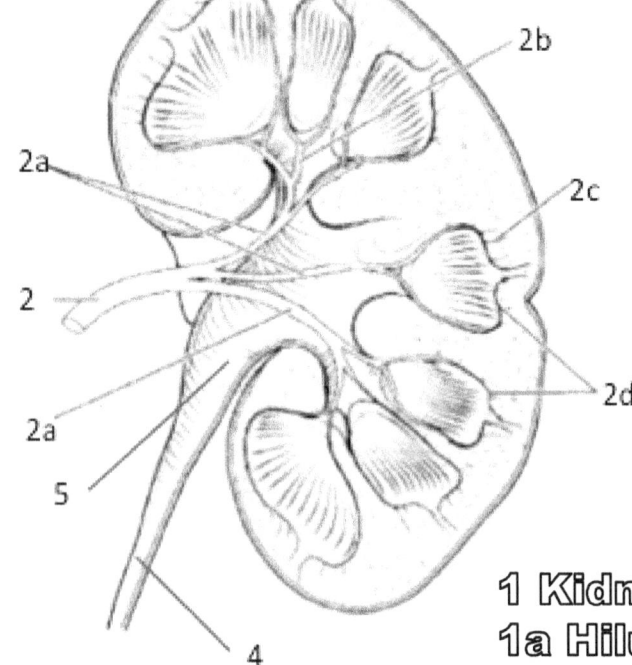

1 Kidney
1a Hilum of Kidney
2 Renal Artery
2a Segmental arteries
2b Interlobar arteries
2c Interlobular arteries
2d Arcuate arteries
3 Renal Vein
4 Ureter
5 Renal Pelvis
6 Renal Capsule
7a Renal Cortex
7b Renal Medulla
7c Renal Column
8a Minor Calyx
8b Major Calyx
9 Renal Pyramid

Excretory 3- Nephron Anatomy

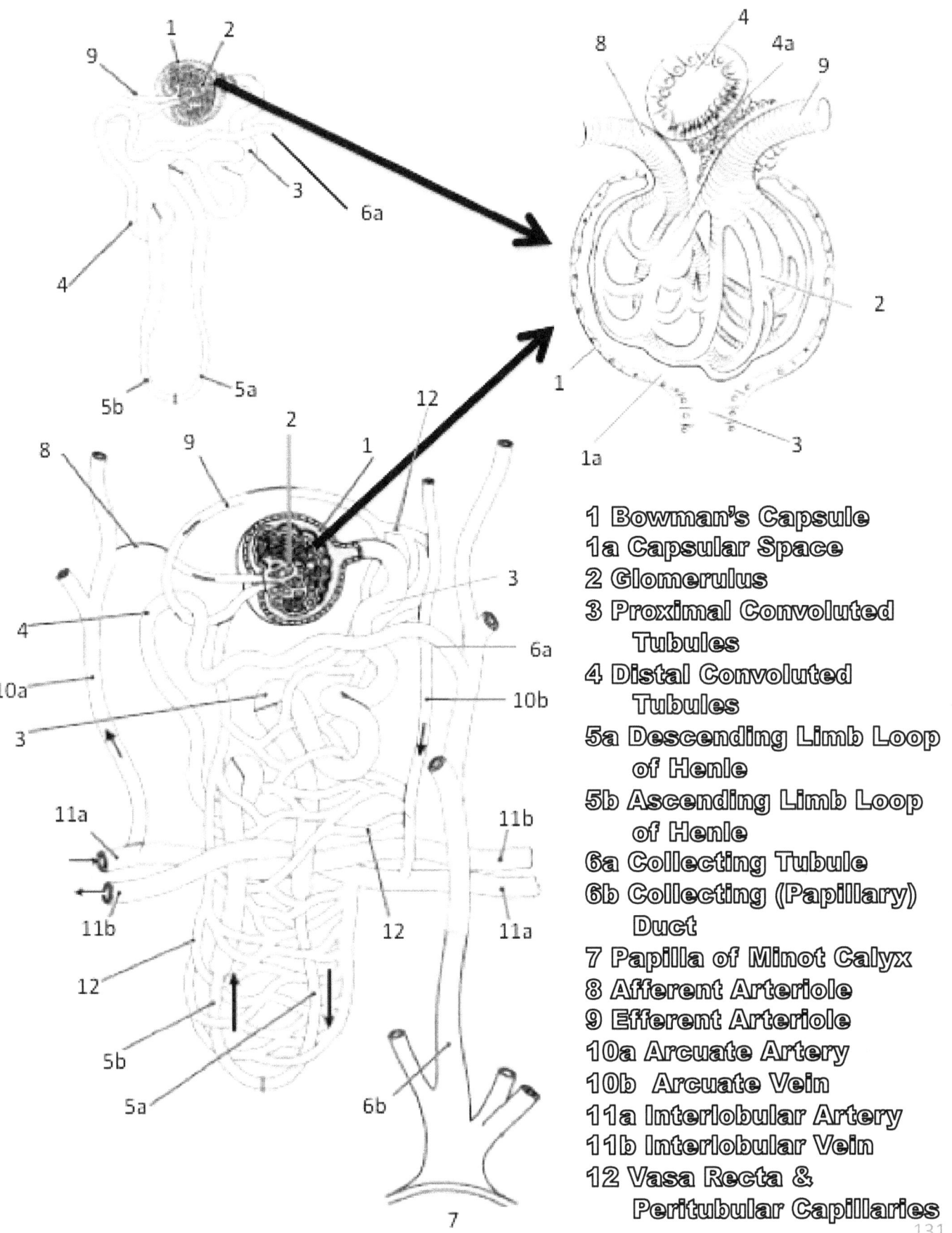

1 Bowman's Capsule
1a Capsular Space
2 Glomerulus
3 Proximal Convoluted Tubules
4 Distal Convoluted Tubules
5a Descending Limb Loop of Henle
5b Ascending Limb Loop of Henle
6a Collecting Tubule
6b Collecting (Papillary) Duct
7 Papilla of Minot Calyx
8 Afferent Arteriole
9 Efferent Arteriole
10a Arcuate Artery
10b Arcuate Vein
11a Interlobular Artery
11b Interlobular Vein
12 Vasa Recta & Peritubular Capillaries

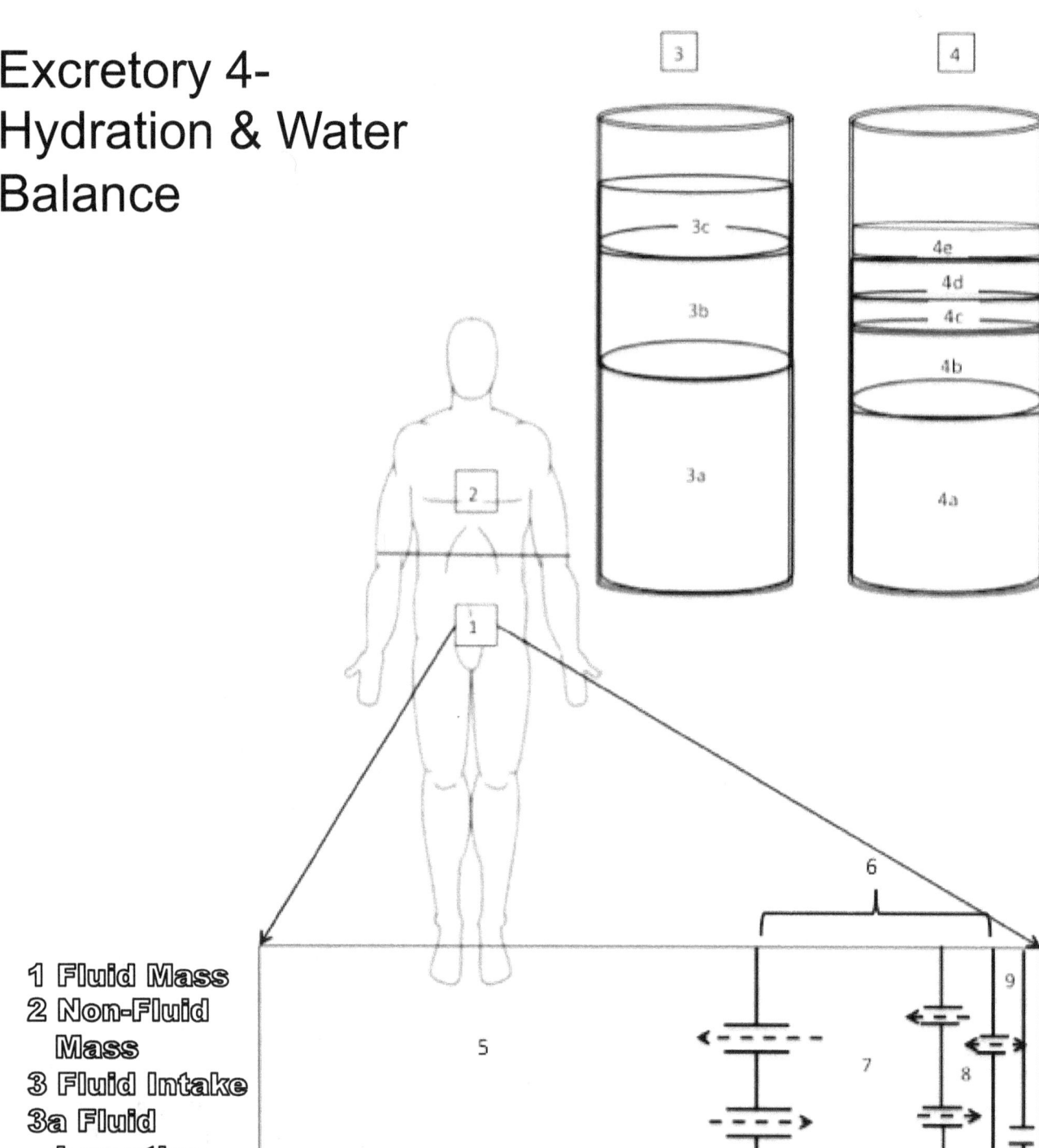

Excretory 4 - Hydration & Water Balance

1 Fluid Mass
2 Non-Fluid Mass
3 Fluid Intake
3a Fluid Ingestion
3b Food Materials
3c Metabolism
4 Fluid Loss
4a Urine Loss
4b Unregulated Loss
4c Sweat Loss
4d GI and Feces Loss
5 Intracellular Fluid (63% of total Water)
6 Extracellular (37% of total Water)
7 Interstital Fluid
8 Plasma
9 Transcellular (Lymph, Secretions, CSF)

Reproductive System & Development

Overview of the System

The reproductive system is a **DIMORPHIC SYSTEM THAT IS "GENDER SPECIFIC"** and has differential functions for each gender. System begins to develop around week 5-6 as prototypical gonads and then when a spike of Testosterone occurs, male gonads begin to develop and differential morphology is seen at 10 weeks. While, within the developing female the initial cell division of the gametes are taking place leading to the development of primordial follicles.

Functionality of system doesn't occur until **PUBERTY** when a **SPIKE IN TESTOSTERONE LEADS TO MALE DEVELOPMENT AND ESTROGEN/ PROGESTERONE LEADS TO FEMALE DEVELOPMENT**. For the **FEMALE,** the **DEVELOPMENT AND ACQUISITION OF ORNAMENTAL (SUBCUTANEOUS) FAT DEPOSITS** are needed for developing a functional reproductive system that also is associated with **MUSCULOSKELETAL ADJUSTMENTS WITHIN PELVIC BOWEL NECESSARY FOR BIRTHING.** While both morphologies have the primarily responsibility for the generation of gametes and structures that allow for the fusion of gametes for the production of viable offspring, each provides different stimulus for morphology changes throughout the lifespan of the individual.

The **PRINCIPLE ORGANS OF THE FEMALE REPRODUCTIVE SYSTEM ARE THE UTERUS, OVARIES, EXTERNAL GENITALS, AND MAMMARY GLANDS**. System provides the organs and secretions that provide protection and support during gestation (the growth, development and maturation) of the offspring. The **PRINCIPLE ORGANS OF THE MALE REPRODUCTIVE SYSTEM ARE THE TESTES, PENIS, AND PROSTATE GLAND.** Where regardless of the individual all hypothalamic regulation is based on estradiol feedback on gonadotropic cells of the hypothalamus, with for males some involvement with testosterone as well. These gonadotropic cells are principally found in the Anterior, Lateral, Paraventricular, Pre-optic, Ventromedial Nuclei and Mammary Body of the hypothalamus. These same nuclei are also sensitive to thermoreceptors, which is one of the rationales for "hot flash" issues females have during the menopause sequence or during hormone replacement therapy. All of which also have involvement with psychological and somatic responses and modulation of oxytocin and prolactin production & release from pituitary along with a regulation and modulation from Agouti-Related Protein, NPY and Leptin via the regulated release of Kisspeptin that modulates GnRH release.

Reproduction will lead to fertilization and then gestation of the fetus. **GESTATION** is the period of offspring development that includes the **STAGES OF (FERTILIZATION and FORMATION OF ZYGOTE, EMBRYOGENESIS AND NEURULATION, MORPHOGENEIS and MATURATION OF THE FETUS AND BIRTHING)** during which the offspring not only grows (from single cell to billions and billions of cells) but also elaborates and undergoes cell specialization in both structure and function in the offspring.

In addition reproduction, deals with the **IDEAS OF GENETICS AND INHERITANCE OF TRAITS** from parents by the offspring child that is best described by reviewing the **MENDELIAN LAWS OF GENETIC INHERITANCE** and issues of genetic modifications that influence the development of the offspring both during gametogenesis and following fertilization (during the period of gestation). Beyond these laws of inheritance, there are issues during development that can influence the morphological development of the offspring that are known as **TERATOGENS**. Lastly, over the last decades we have built a better understanding regarding the issues of what a female (and male) can, or cannot, do that influence the development of functional gametes, fetal development and health of the offspring following gestation.

Reproductive 1a- Male System & Testicle Anatomy

Anatomy of System

1 Penis
1a Glans Penis
1b Head of Penis
2 Testicle
3 Seminiferous Tubule
4 Scrotum
5 Epididymus
6 vas Defrens
6a Ampulla of Seminal Vesicles
6b Ejaculatory Duct
7 Corpus Cavernosum
8 Corpus Spongiosum
9 Urethra
10 Seminal Vesicles
11 Prostate Gland
12 Cowper's (Bulbourethral) Gland
13 Urinary Bladder
14 Pubic Bone
14a Suspensitory Ligament
15a Dorsal Penile Vein
15b Dorsal Penile Artery
15c Penile Nerve
16a Tunica Albuginea
16b Buck's Fascia
16c Areolar Tissue

Testicle & Sperm Anatomy

1 Testis
1a Spermatocord
1b Gonadal Vein
1c Gonadal Artery
1d Nerve
2 Epididymus
2a Efferent Duct
2b Epididymal Duct
3 vas Defrens
4 Lobule of Testis
4a Septum of Lobule
4b Rete Testis
5 Capsule
6 Seminiferous Tubule
7 Basal Membrane of Seminiferous Tubule
8 Spermatogonim
9 Primary Spermatocyte
10 Secondary Spermatocyte
11a Spermatid
11b Sperm
12a Nurse Cells
12b Interstitial Cells (Sertoli & Leydig)
13a Head of Sperm
13b Midpiece of Sperm
13c Tail of Sperm
14a Acrosomes
14b Nucleus
15 Mitochondria
16 Flagellum

Reproductive 1b- Organs of Male System

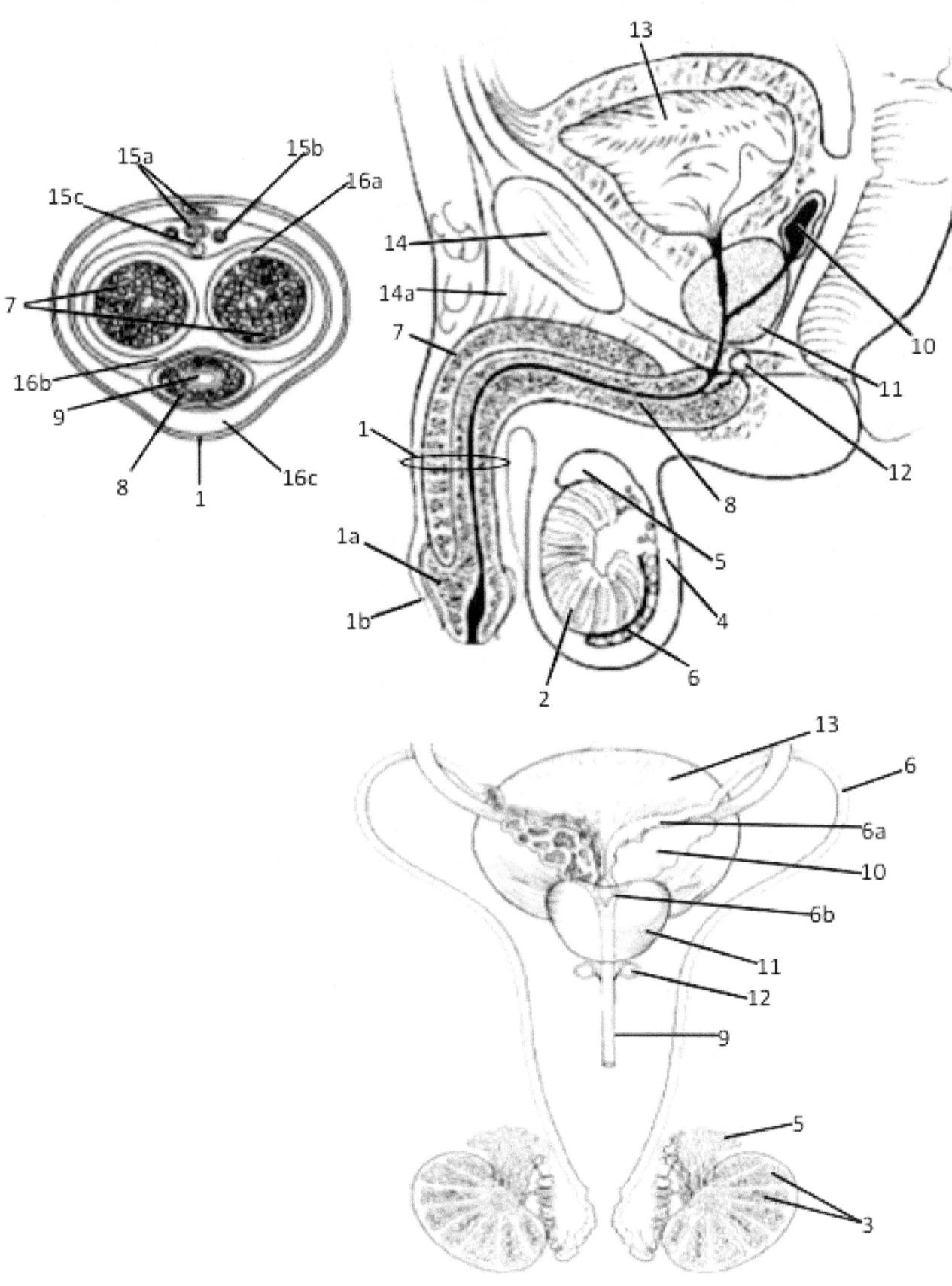

Reproductive 1c- Testicle and Sperm

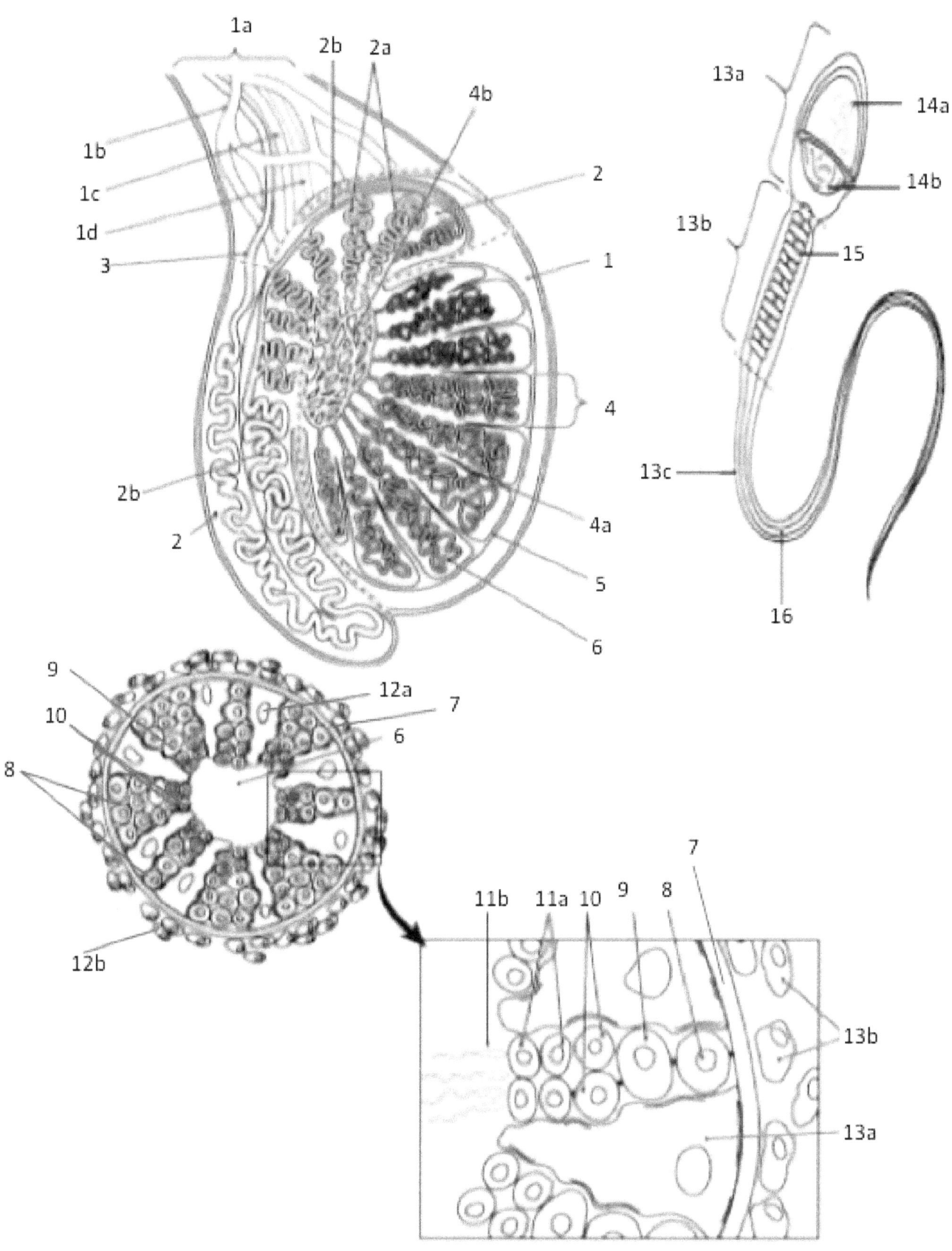

Reproductive 2- Organs of Female System

Urogenital

1 Vagina
1a Vaginal Orifice
1b Skene's Gland
2a Labia Minora
2b Labia Majora
3 Clitoris
4 Mons Pubis
5 Vestibule
6 Cervix
6a Fornix
7 Fimbriae & Fallopian Tube
8 Ovary
9a Round Ligament
9b Uterosacral Ligament
10 Uterus
11 Pubic Bone
12 Urinary Bladder
12a Urethra
12b Anal Triangle
12c Urogenital Triangle
13 Colon
13a Anus

Mammary Gland

14 Mammary Gland (Lobes)
14a Lobular Duct
15 Lacteous Sinus
15a Nipple
15b Areola
16 Blood Vessel
17 Adipose Tissue
18 Costals
18a Pectoralis Major

Reproductive 2a- Organs of Female System

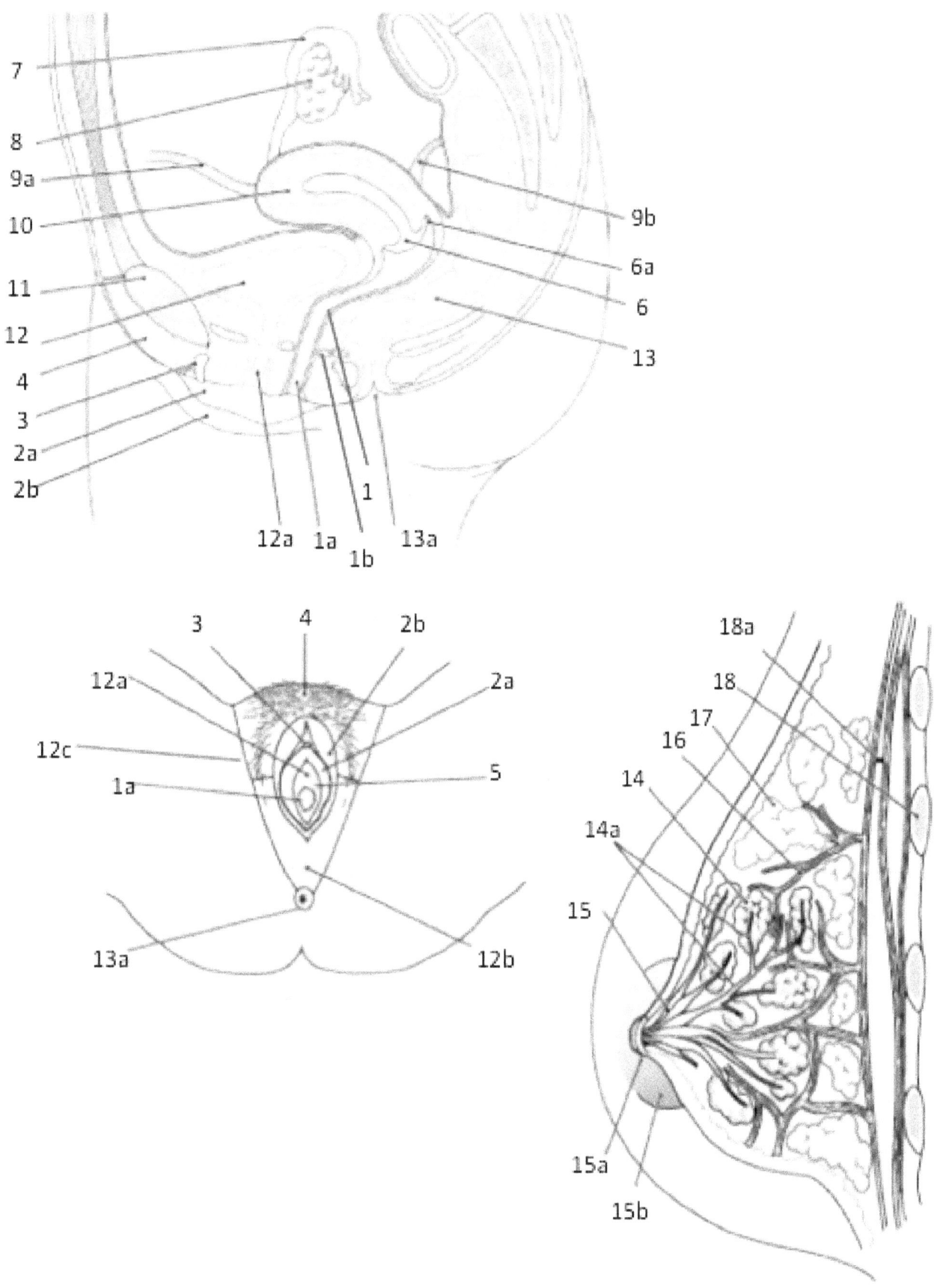

Reproductive 3- Uterus & Vaginal Anatomy, Ovulation and Regulation

Ovulation & Regulation

1 Follicular Phase
1a Menstruation of Endometrium
1b Endometrial Proliferation Phase
2 Ovulation
2b Endometrial Secretory Phase
3 Luteal Phase
3a Endometrial Sloughing Phase

Uterus & Vagina Anatomy

4a Fundus of Uterus
4b Body of Uterus
4c Isthmus of Uterus
4d Cervix of Uterus
4e Endometrial Gland
5a Fimbriae
5b Fallopian/Uterine Tube
6 Ovary
7 Primordial Follicle
8 Primary Follicle
9a Mature Follicle
9b Secondary Oocyte
10 Primary Follicle
11 Ovulate
12a Corpus Luteum
12b Corpus Albicans
13a Ovarian Ligament
13b Ovarian Suspensitory Ligament
14 Broad Ligament
14b Uterosacral Ligament
15 Vagina
15 Vaginal Orifice
16 Rugae of Vagina

Reproductive 3a- Uterus & Vaginal Anatomy, Ovulation and Regulation

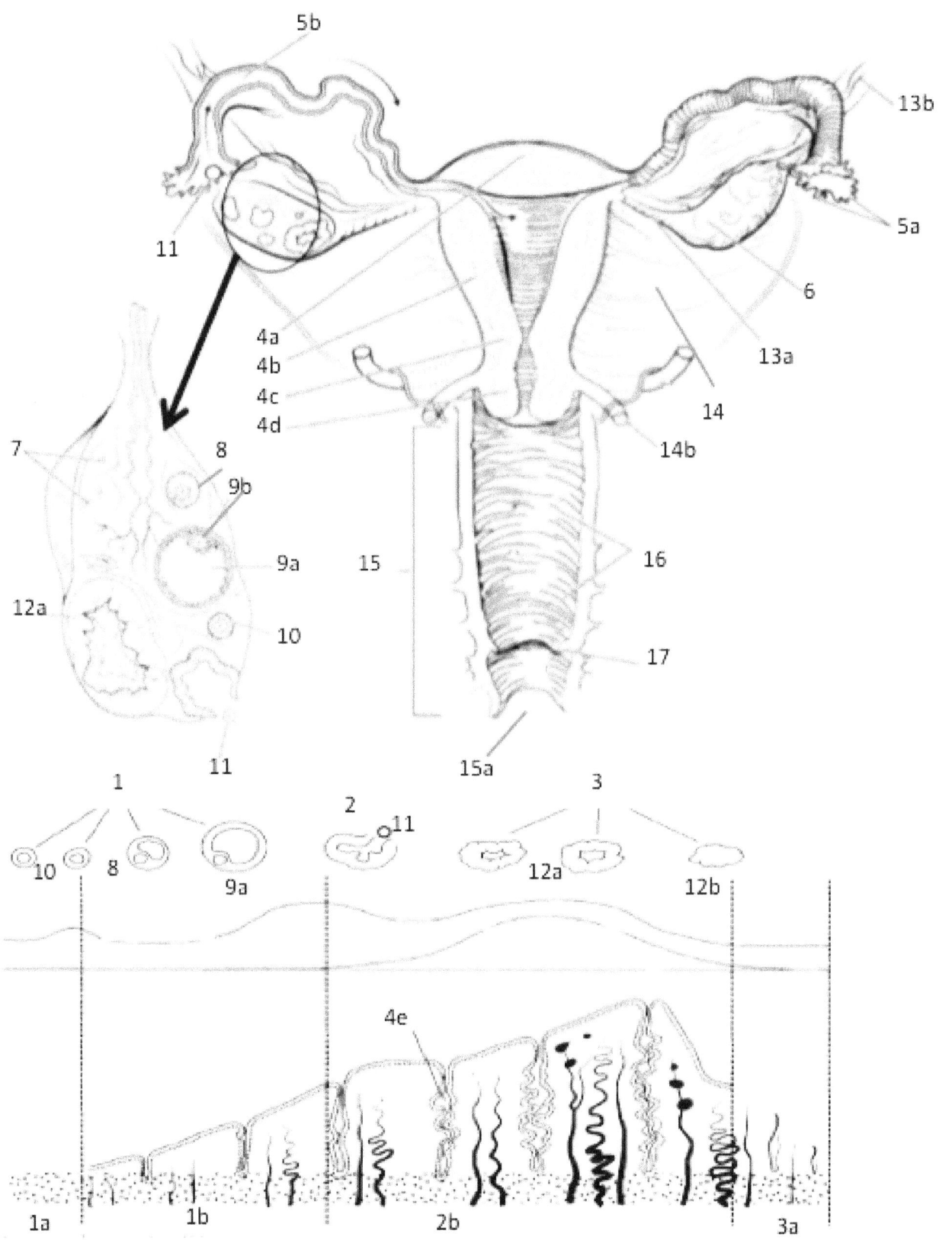

Reproductive 5a- Meiosis

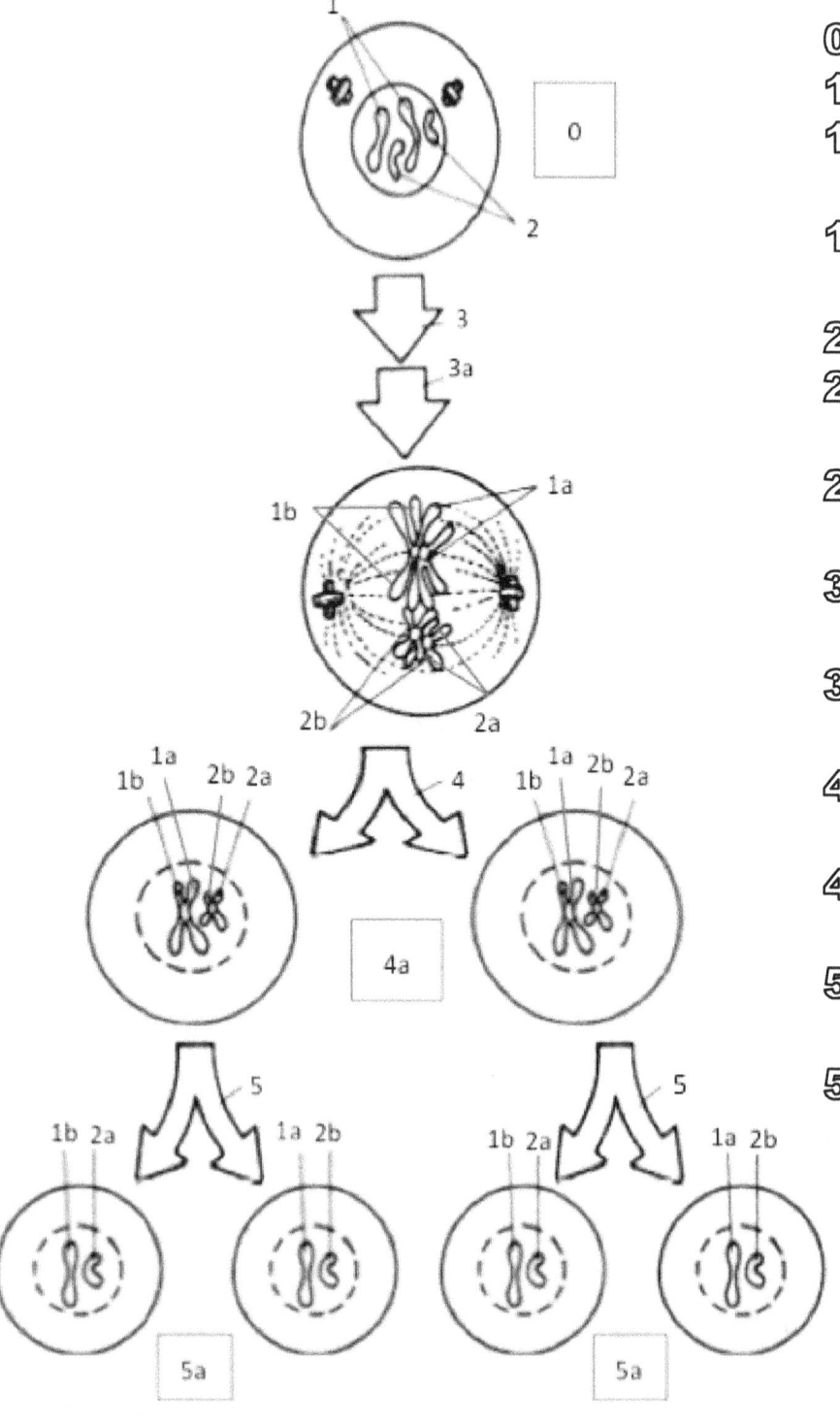

0 Stem Cell (2N)
1 Chromosome A
1a Homologous Chromatid 1 A
1b Homologous Chromatid 2 A
2 Chromosome B
2a Homologous Chromatid 1 B
2b Homologous Chromatid 2 B
3 Growth and Prophase
3a Crossing-over Event
4 Anaphase & Telophase
4a Daughter Cells (2N)
5 Cell division without growth
5a Gametes (N)

Reproductive 5b- Gametogenesis

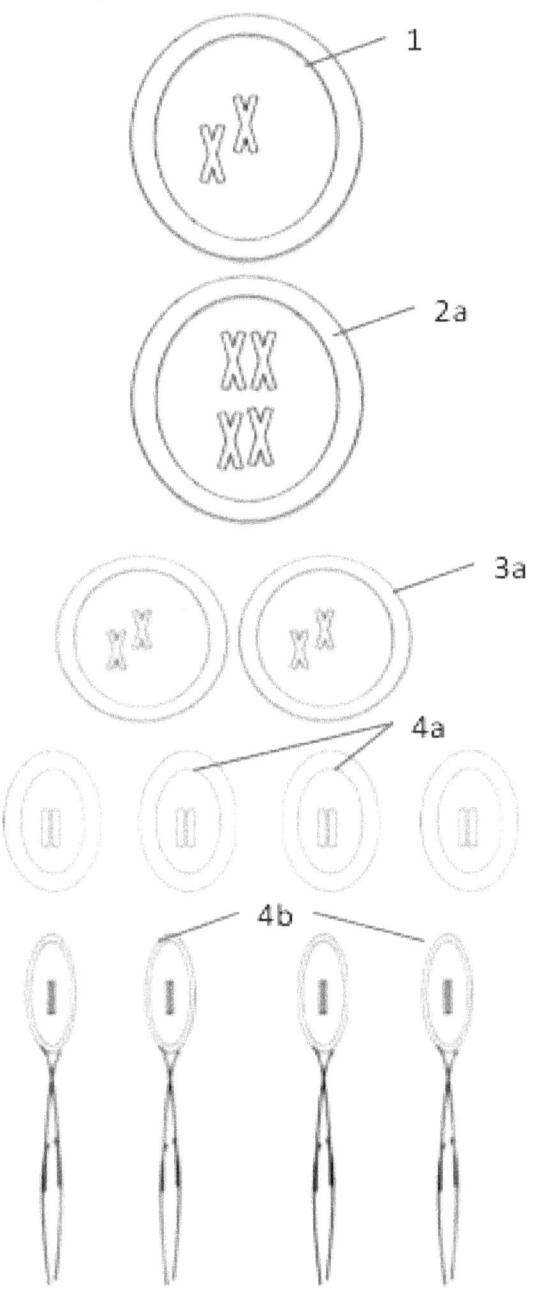
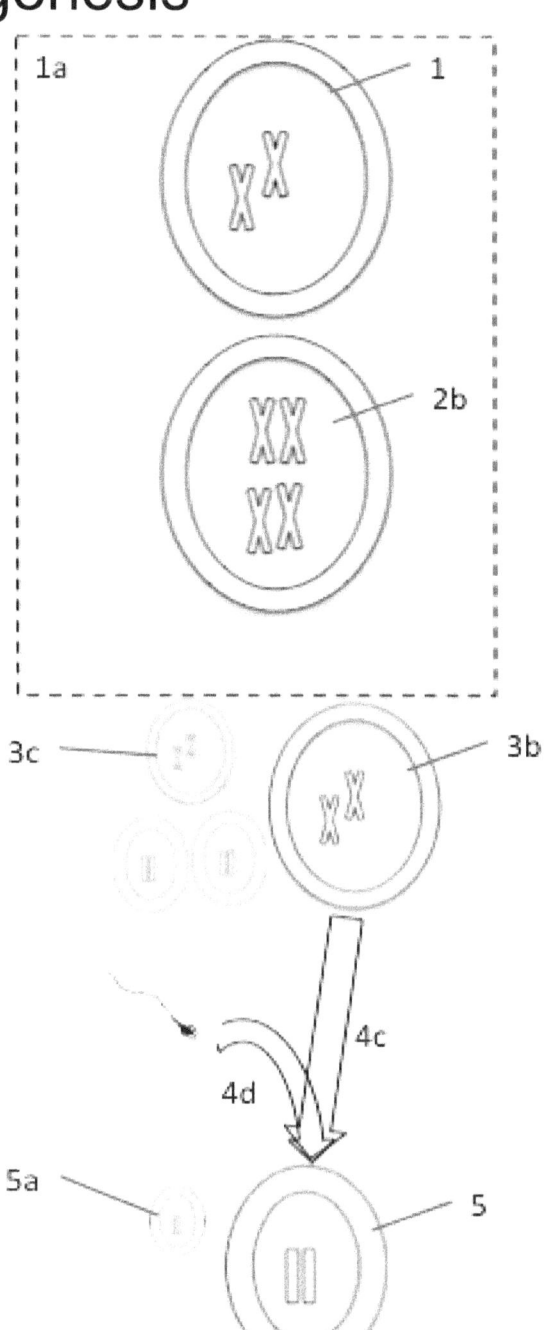

1 Germal Stem Cell
1a During gestation
2a Spermatogonia
2b Oogonia
3a Primary Spermatocyte
3b Primary Oocyte
3c Primary Polar Bodies

4a Secondary Spermatocyte
4b Sperm
4c Ovulation
4d Sperm interaction
5 Secondary Oocyte
5a Secondary Polar Body

Reproductive 6a- Fertilization and Implantation

1 Ovulation
1a Oocyte
1b Zona Pellucida
1c Corona Radiate
2 Fertilization
2a Sperm
3 Cleavage
3a Binary Fission
4 Blastulation
4a Blastocyst
5 Area of Implantation
5a Trophoblast (Becomes Placenta)
5b Inner Cell Mass (Becomes Embryo)
5c Blastocoel

6a Amniotic Cavity
6b Embryonic Disc
6c Fetus
7 Chorion
7a Chorionic Cavity
8 Endometrium
9 Uterus
9a Uterine Cavity

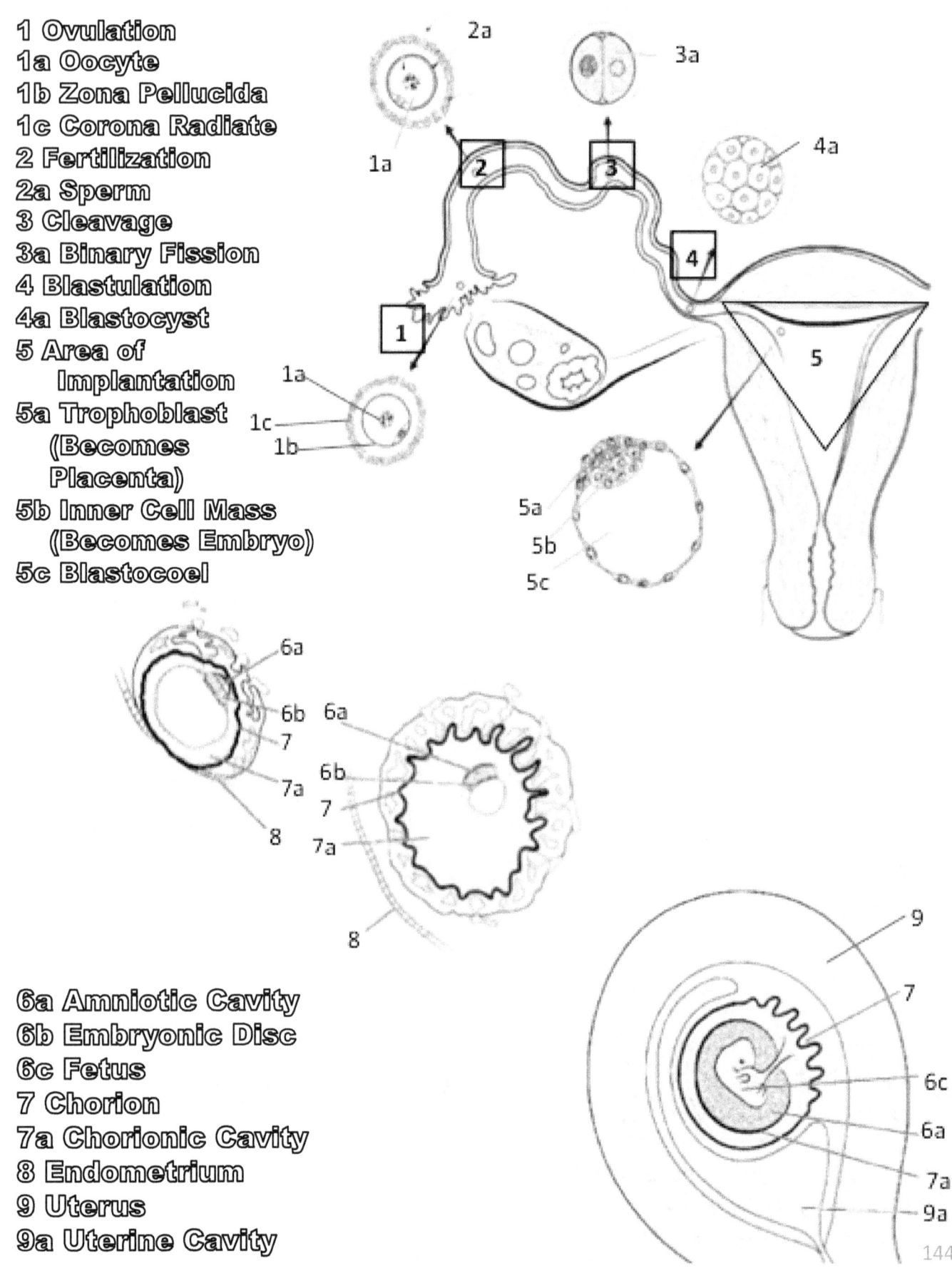

Reproductive 6b- Embryogenesis & Gastrulation

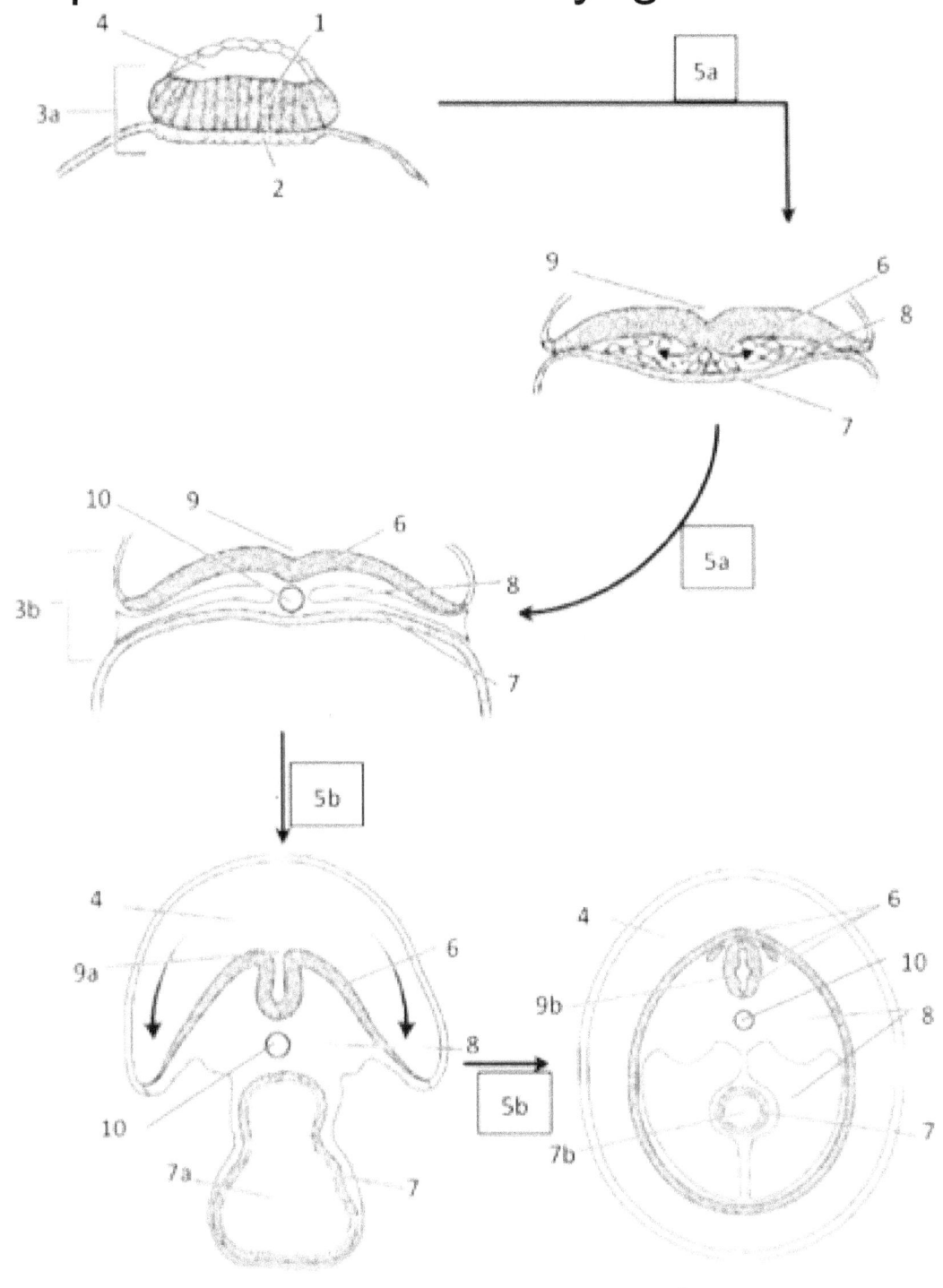

1 Epiblast
2 Hypoblast
3a Germal Bilayer
3b Germal Trilayer
4 Amniotic Cavity
5a Gastrulation
5b Neurulation
6 Ectoderm
7 Endoderm
7a Yolk Sac
7b Gut
8 Mesoderm
9 Primitive Streak
9a Neural Groove
9b Neural Tube
10 Notochord

Reproductive 9- Fetal Circulation

1 Uterus
1a Uterine Arteries
1b Uterine Vein
2 Fetal Chorion
2a Placental Capillaries
2b Umbilical Vein
2c Umbilical Artery
4 Chorion Venal Sinus
4a Chorionic Villi
5a Ductus Venous
5b Ductus Arteriosus
6 Inferior Vena Cava
7 Aorta
7a Aortic Arch
8 Pulmonary Trunk
9 Iliac Artery
10a Aorta of Heart
10b Ventricle of Heart
11 Liver

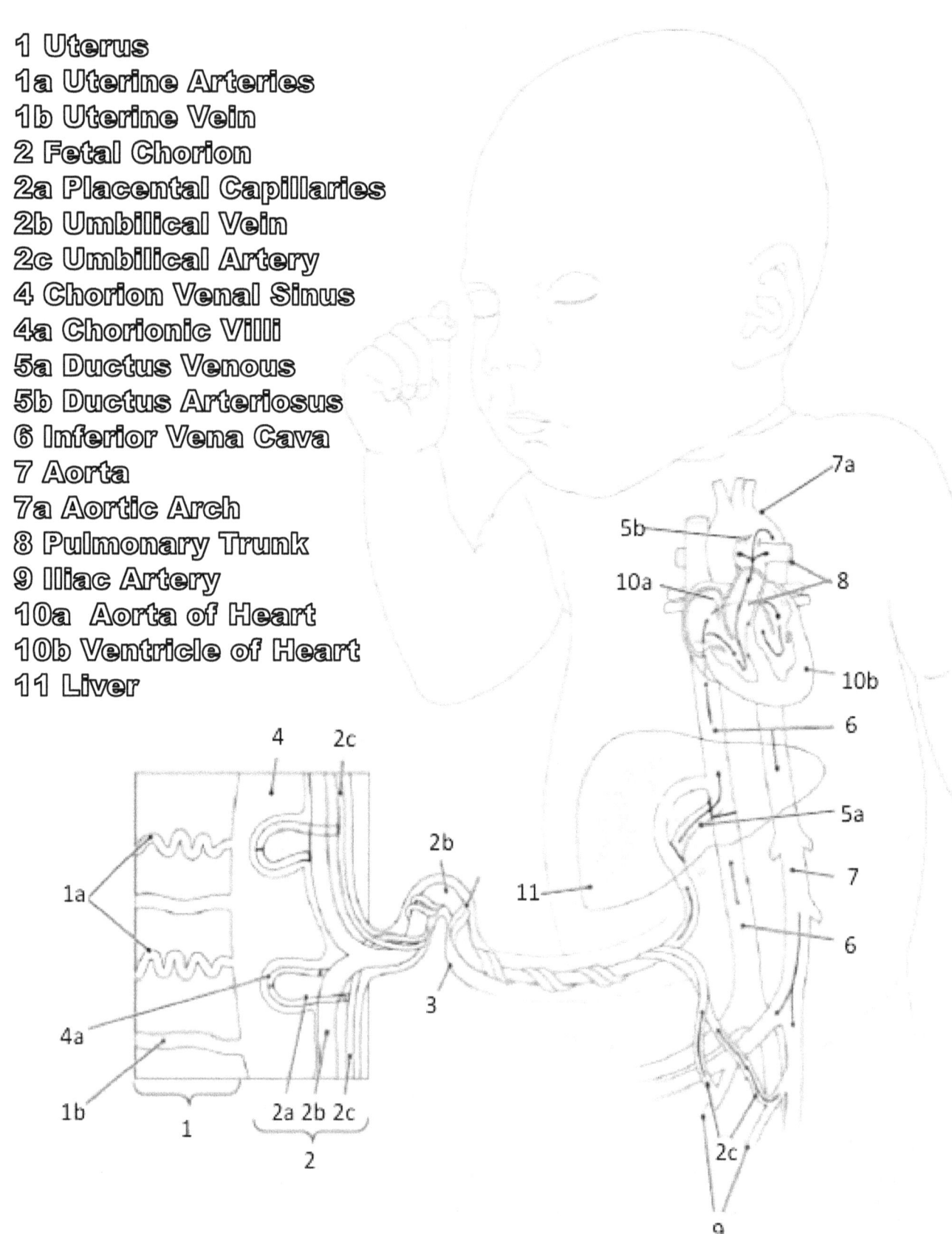

Reproductive 10-Summary of Development

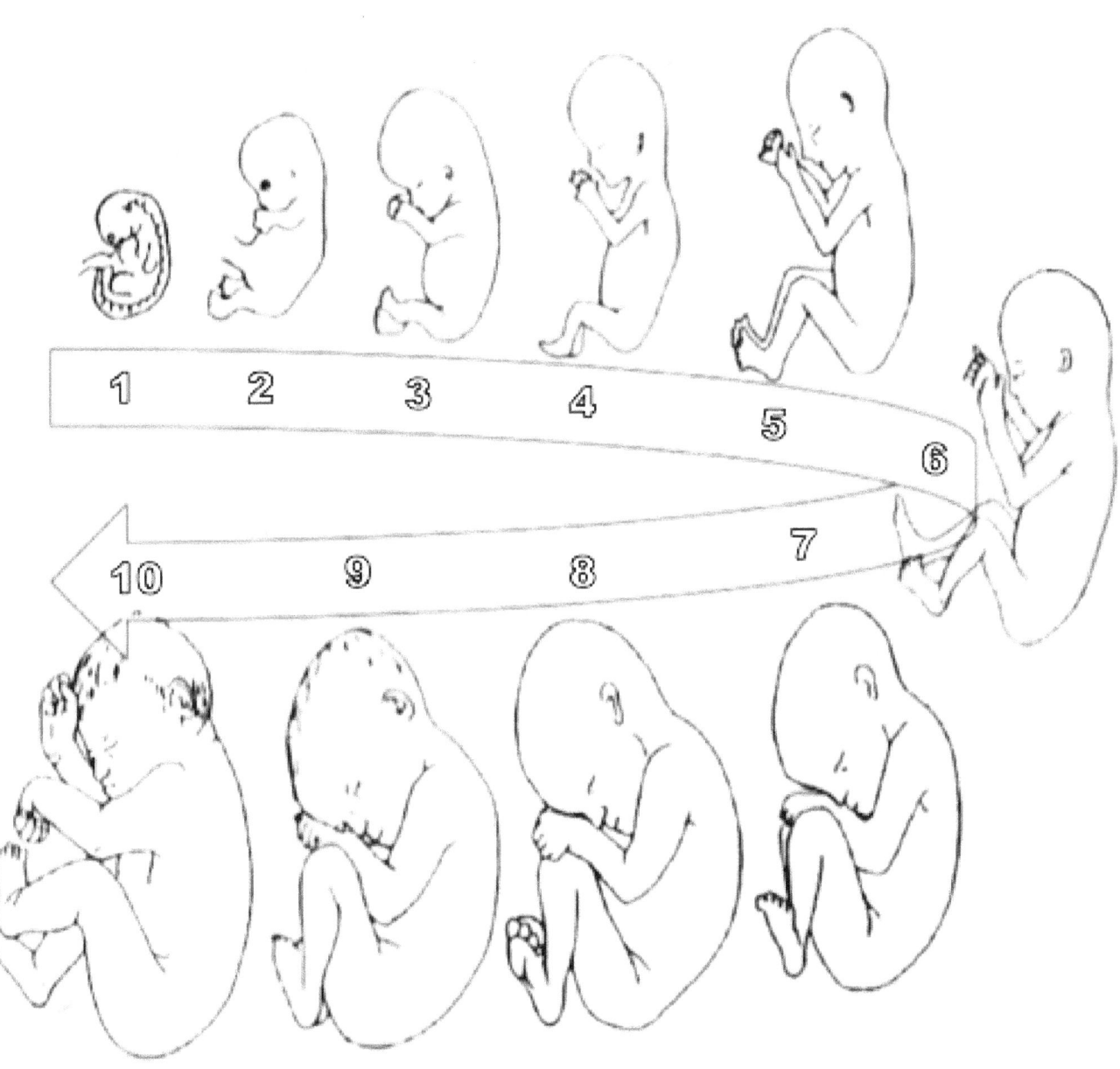

1 @ 4 weeks 4-7 mm, < 1g
2 @ 8 weeks 14-20 mm, 1-2 g
3 @ 12 weeks 75-85 mm, 14 g
4 @ 16 weeks 160 mm, 100 g
5 @ 20 weeks 240 mm, 300 g
6 @ 24 weeks 300 mm, 600 g
7 @ 28 weeks 350 mm, 1000 g
8 @ 32 weeks 410 mm, 1700 g
9 @ 36 weeks 460 mm, 2400 g
10 @ Full-term 490-500 mm, 3090 g

www.ingramcontent.com/pod-product-compliance
Lightning Source LLC
Chambersburg PA
CBHW082204220526
45470CB00010B/3036